Exploring Extended Realities

This volume highlights interdisciplinary research on the ethical, metaphysical, and experimental dimensions of extended reality technologies, including virtual and augmented realities. It explores themes connected to the nature of virtual objects, the value of virtual experiences and relationships, experimental ethics, moral psychology in the metaverse, and game/simulation design.

Extended reality (XR) refers to a family of technologies aiming to augment (AR) or virtually replace (VR) human experience. The chapters in this volume represent cutting-edge research on XR experiences from a wide range of approaches including philosophy, psychology, Africana studies, and cognitive sciences. They are organized around three guiding questions. Part 1, "What is Extended Reality?", contains a series of chapters examining metaphysical questions about virtual objects, actions, and worlds. Part 2, "Is There an Ethics for Extended Realities?", includes chapters that address ethical questions that arise within XR experiences. Finally, Part 3, "What Can We Do with Extended Realities?", features chapters from a diverse group of social scientists on the potential uses of XR as an investigative and educational tool, including its strengths and pitfalls.

Exploring Extended Realities will appeal to scholars and advanced students working in philosophy of technology, metaphysics, moral psychology, applied ethics, and game studies.

Andrew Kissel is Assistant Professor of Philosophy and Religious Studies at Old Dominion University. His research focuses on the foundations of moral responsibility in virtual and non-virtual contexts. With the Virginia Modelling, Analysis, and Simulation Center (VMASC), he develops thought experiments in VR, including work funded by the National Endowment for the Humanities.

Erick José Ramirez is Associate Professor of Philosophy at Santa Clara University. He has developed numerous thought experiments for virtual reality, which are available on his personal website: https://www.erickjramirez.com. He is the author of *The Ethics of Virtual and Augmented Reality: Building Worlds* (Routledge).

Routledge Studies in Contemporary Philosophy

Exploring Extended Realities

Metaphysical, Psychological,
and Ethical Challenges

Edited by
Andrew Kissel and Erick José Ramirez

Routledge
Taylor & Francis Group

NEW YORK AND LONDON

First published 2024
by Routledge
605 Third Avenue, New York, NY 10158

and by Routledge
4 Park Square, Milton Park, Abingdon, Oxon, OX14 4RN

Routledge is an imprint of the Taylor & Francis Group, an informa business

ISBN: 978-1-032-41732-5 (hbk)
ISBN: 978-1-032-41731-8 (pbk)
ISBN: 978-1-003-35949-4 (ebk)

DOI: 10.4324/9781003359494

Typeset in Sabon
by codeMantra

To friendships
(physical, virtual, and everything in between)

Contents

Contributors

Espen Aarseth is professor of game studies and head of the PhD School at the IT-University of Copenhagen and a member of the Royal Danish Academy of Sciences and Letters and of the Academia Europaea. He is founding editor of the journal *Game Studies*. He is also a special terms professor at the Digital Media Department at Beijing Normal University.

Jeremy N. Bailenson is the founding director of Stanford University's Virtual Human Interaction Lab and Thomas More Storke Professor in the Department of Communication. His research is central to the study of virtual reality, moral psychology, and digital storytelling. He is the author of *Experience on Demand: What Virtual Reality Is, How It Works, and What It Can Do.*

Dorian Clay graduated from Santa Clara University in 2022 with degrees in Computer Science and Engineering along with an Independent Studies degree focusing on Science, Technology, and Ethics. He is currently working as a software engineer for Microsoft.

Kathryn B. Francis is a Lecturer at Keele University. Her research focuses on investigating moral judgment-action discrepancy using VR and moral decisions in applied contexts such as emergency/medical decision-making and diet choice. She is published in psychology-specific journals such as *European Journal of Social Psychology*, general science journals such as *Scientific Reports*, and philosophy journals such as *Episteme* and *Synthese*.

Mohit Gandhi graduated from Santa Clara University in 2021 with degrees in Philosophy and Computer Science. He currently works for AVEVA as an XR Application Engineer in San Francisco, California.

Eugy Han is a Communication PhD student at Stanford University. She earned her B.S. in Cognitive Science from Brown University in 2020. Han's research focuses on how virtual reality (VR) environments and the embodiment of digital identities transform people's behaviors and social interactions.

Shelby Jennett is a student at Santa Clara University. Her research focuses on the intersections of Neuroscience, Philosophy of Technology, and Applied Ethics.

Michael Madary is an Assistant Professor of Philosophy at the University of the Pacific. His main area of teaching and research is philosophy of mind from an interdisciplinary perspective. Recent work includes exploring the experience, and ethics, of virtual reality technologies. He is the author of *Visual Phenomenology*.

Joseph Neisser is Professor of Philosophy and Adjunct Professor to the Program in Neuroscience at Grinnell College, specializing in neurophilosophy and neurophenomenology. He is also the author of *The Science of Subjectivity* (Palgrave Macmillan), as well as several articles and chapters. He can be found on the web at https://neisserj.wordpress.com/.

Thomas D. Parsons is the Grace Center Professor of Innovation in Clinical Education, Simulation Science, and Immersive Technology. He also directs the Computational Neuropsychology and Simulation (CNS) laboratory. His most recent work includes experimental and therapeutic research using virtual reality. He has published five books including *Ethical Challenges in Digital Psychology and Cyberpsychology*.

Jon Rueda is Ph.D. candidate at the University of Granada, Spain. He has published on diverse topics of Ethics of Technology, AI Ethics, VR Ethics, Bioethics, Neuroethics, and Public Health Ethics. He has completed research stays at the Uehiro Centre for Practical Ethics at Oxford University and at the Ethics Institute at Utrecht University.

Mark Silcox is Professor of Philosophy and Chair of the Department of Humanities and Philosophy at the University of Central Oklahoma. His research interests include metaphysical investigations into simulated objects. He is the author of *A Defense of Simulated Experience: New Noble Lies* (Routledge) and editor of *Experience Machines: The Philosophy of Virtual Worlds*.

Grant Tavinor is a Philosopher of the Arts at Lincoln University in New Zealand. His research focuses on digital technology, including videogames and virtual reality, and its philosophical, aesthetic, and ethical dimensions. He is the author of *The Aesthetics of Virtual Reality* (Routledge) and *The Art of Videogames*.

Acknowledgements

This collection would not exist (physically, digitally, or otherwise) without the help of many others. First and foremost, we would like to jointly thank our friend and colleague Justin Remhof (Old Dominion University) who first introduced the two of us and which started us off on several fruitful collaborations. Thank you, Justin! The authors would also like to thank Andrew Weckenmann and Rosaleah Stammler at Routledge for their guidance and help and for trusting in us as we collaborated on this collection. They were, truly, great to work with, responsive, and always ready with good advice.

A collection of articles is nothing without the work of the scholars who contributed to it. We thank them all for bringing exciting new ideas on the metaverse into the world. Each of the authors in this volume contributes something unique, and it was an honor to read through and comment on their work. In our role as editors and philosophers, we learned a lot from all of you, and we hope that readers will feel the same way as they wrestle with your ideas.

Kissel's path to this book started with a little course I taught on philosophy and videogames. Thank you to Dylan Wittkower and Yvette Pearson for their help and encouragement making that happen. I would also like to thank Kevin Moberly, Marc Ouellette, and all the wonderful folks at the annual SWPACA conference for introducing me to game studies and for their continued helpful advice and feedback as I explore philosophy in virtual worlds. You've pulled me deeper into the metaverse.

Thank you to Krsyztof Rechowicz and John Shull at the Virginia Modelling, Analysis, and Simulation Center. They've opened my eyes to the joys and pains of creating XR experiences. Without them, I never would have been able to pursue these projects. And thank you to the National Endowment for the Humanities for taking a gamble on research led by a young philosopher. There were numerous important folks that tested, broke, and gave tips for how to fix my work, including David Schwan, Mohammad Obeid, and Jeff Langevin. And thank you all for agreeing to continue down this path with me, along with newcomers Megan Mize and Ian Quitadamo.

Conversations with panels, reading groups, colleagues, and even focus groups have shaped my thinking on these topics. Thank you especially to Ashley Shew and the good folks at SEPOT, Melissa McDonald, Kathryn Francis, Justin Remhoff, Lauren Eichler, Andrew Garner, Chris Tweedt, Torrance Fung, Nicole Willock, and Andrew Gould. Some of you are avid VR-heads and others I couldn't pay to put on an HMD, but all of you provided invaluable perspectives in shaping my work. I apologize to all of the folks that I cannot include here due to space and the poor functioning of my memory, but thank you from the bottom of my heart.

I have the joy and privilege of having a family who not only supports me emotionally and mentally but are also willing to challenge my harebrained philosophical views. Thank you to all of them, but especially to my wife, Teresa. Someday I hope to be half the philosopher that she is.

Finally, thank you to my co-editor, colleague, and friend Erick Ramirez. I'm so glad that Justin encouraged us to reach out to each other. Working with you has been a privilege and an honor. I'm sorry to see the end of this project, as I no longer have the excuse to Zoom-call you to chat about philosophy, obscure movies from the 90s, and videogames, but I look forward to future collaborations. Perhaps someday we'll meet in the flesh, but for now, I'm happy to count you among my dearest virtual friends.

Ramirez would like to also personally thank Andrew Kissel for his tireless work and the genuine spirit of camaraderie he brought to all of our conversations and meetings, including the ones where we got actual work done. I've never met Kissel in physical space, but I count him as a good friend. I also want to thank Susan Kennedy at Santa Clara University. Susan is a philosopher of technology whose work I both admire and I've learned a lot from. Our conversations have helped shape my thoughts on many issues. I also want to thank my co-authors Shelby Jennett, Dorian Clay, and Mohit Gandhi. It has been wonderful (and great fun) working with you on this project and developing ideas together.

Brian Green, Irina Raicu, and Don Heider at the Markkula Center for Applied Ethics have all supported me intellectually, personally, and financially on projects that directly and indirectly informed the ideas in my contribution. They're great people at a great center. Lastly, I want to thank my partner Maggie Levantovskaya. She's both a writer and a fellow academic. She's awesome in too many ways to list and she's helped me become a better writer, philosopher, and person. Much love Maggie <3

Introducing Exploring Extended Realities
Metaphysical, Psychological, and Ethical Challenges

Andrew Kissel and Erick José Ramirez

In 2020, Facebook (now Meta) released the Oculus Quest 2, the best-selling virtual reality (VR) headset of all time. Boosted by the COVID-19 pandemic, the Quest 2 and a variety of other head-mounted displays (HMDs) transformed extended reality (XR) technologies from the realm of science fiction and futurism into a household regularity. And yet, many thinkers reflecting on the philosophical, metaphysical, and ethical questions raised by these technologies have clung reticently to a fantastically science-fictional view of XR as giving rise to wholly realized virtual worlds that are indistinguishable from their non-virtual counterparts. This is perhaps unsurprising. From the Upanishads to Zhuangzi to Descartes and David Chalmers, innumerable thinkers have taken up the possibility that the world as it presents itself to us masks an alternative reality. XR, on this view, is a contemporary partial realization of the kinds of all-encompassing thought experiments that have cropped up throughout the history of thought.

Perhaps another cause of this attitude is a tendency to focus on one extreme of the XR spectrum. In this collection, unless otherwise noted, we use the term XR to refer broadly to the spectrum of digital technologies that create and/or modify partial or total virtual environments in a multidimensional space (Vi, da Silva, & Maurer 2019). At one end of this spectrum are VR technologies, which attempt to block the physical world entirely and fully immerse users in virtual environments. Most uses of the Quest 2 HMD would count as VR in this sense. At the other end of the XR spectrum are augmented reality (AR) technologies, which overlay virtual elements on the physical world. The popular Pokemon GO (Niantic 2016) app, which allows players to use their camera phones to view virtual monsters overlaid in their physical environments, is an example of AR. In between these extremes are a variety of technologies that vary in the degree to which the virtual environment replaces or overlays the surrounding physical environment.

DOI: 10.4324/9781003359494-1

The boundaries along this spectrum are vague. For example, many VR technologies, including the Quest 2, allow for *pass-through* modes, wherein externally facing cameras allow video of the external physical environment to "pass through" into an otherwise self-contained VR environment (usually to prevent knocking into physical objects that would otherwise not be seen by the user). Furthermore, the kinds of sensory input that technologies along this spectrum will provide vary widely. Some, such as Google's Cardboard VR, involve placing the user's smartphone in a cardboard visor to create a simple HMD. The resultant experience provides a 3D visual environment that is responsive to the user's head movements, and perhaps some audio feedback. However, it lacks other forms of sensory input and response, such as bodily motion tracking, haptic feedback, olfactory feedback, and the like. STEPVR's Gates01, in contrast, uses a massive metal frame to integrate an HMD, an omnidirectional treadmill, and a haptic feedback vest (among other sensory technologies) in an attempt to create a more immersive experience for the user. And yet, experiences on both Cardboard VR and Gates01 could count as falling on the VR side of the XR spectrum.

In many cases, those using and developing XR technologies are interested in two related but distinct goals: immersion and presence. The use of these two terms is debated and varies slightly across the literature (Nilsson et al. 2016; Weber et al. 2021). Nevertheless, as a general claim, *presence* refers to the subjective psychological experience of location-illusion, the experience of "being-there" in the virtual environment. *Immersion*, by contrast, is defined in terms of the hardware and software properties that help generate presence. The distinction as drawn here corresponds to Cummings and Bailenson (2016) when they write:

> [P]resence in a [virtual environment] is inherently a function of the user's psychology, representing the extent to which an individual experiences the virtual setting as the one in which they are consciously present. On the other hand, immersion can be regarded as a quality of the system's technology, an objective measure of the extent to which the system presents a vivid virtual environment while shutting out physical reality.
>
> (274)

The felt experience of presence has been leveraged for a variety of purposes, including gaming, training simulations, exposure therapy, education, research, and even customizing sales for makeup, glasses, and cars, to name a few. Many of these uses and more are discussed and represented in this volume.

As can be seen, then, the technologies that fall on the XR spectrum are wide and varied, as are their uses. The metaphysical, psychological, and ethical questions that they give rise to vary as well. AR experiences like Pokemon GO do not obviously give rise to the kinds of all-encompassing skeptical worries that have historically been the bread and butter of discussions of the philosophical problems that arise from immersive VR. The essays in this collection reflect this broad variation in form and use of XR technologies. Similarly, the authors represent diverse approaches to the questions XR gives rise to, including philosophy, psychology, neuroscience, and game studies. The topics are by no means exhaustive, but all share a commitment to engaging with XR technologies as they are now or are immediately headed.

So what is on the horizon when we explore XR? The collection is broadly divided into three sections, each motivated by a particular question. Those questions are roughly, "What is it?", "What should we do *in* it?", and "What should we do *with* it?" Of course, the division into sections is slightly artificial. As will become obvious to the reader as they work through each of the chapters, questions about the metaphysics of XR are deeply intertwined with ethical questions, both of which have shaped and informed the way that XR affects those who create, distribute, and use XR experiences. Central themes, such as questions of the self and identity, embodiment, and what counts as real, resonate in all three sections. In what follows, we lay out the papers and identify some of the key themes that occur and re-occur.

"What Is Extended Reality?"

Are virtual objects, actions, and events *real*, or are they merely useful fictions? In his highly publicized work *Reality +: Virtual Worlds and the Problems of Philosophy*, David Chalmers defends "virtual realism", the thesis that virtual realities are genuine realities (Chalmers 2022, 106). He contrasts this with "virtual fictionalism", the view that virtual realities are fictional realities (193). At stake is the meaning of what we are doing when we enter XR. Are we entering potentially fruitful, but ultimately fictional, worlds that create compelling illusions? Or are we entering a world whose ontology, though perhaps different in some respects from its non-virtual counterparts, should be considered among the real existent things? The first three chapters in this section focus on the debate between the virtual fictionalist and the virtual realist, each taking a different stance.

The first chapter from Espen Aarseth argues that his and others' earlier work in game studies largely anticipates Chalmers' position. As such, he agrees that virtual objects are real: they're virtually real. Nevertheless, he argues that placing realism and fictionalism as competing alternatives (and lumping them in with the digital technologies they often run on) misses the complexity of virtual objects. He traces much of the difficulties of the debate to differences in how we understand the term "fiction" in the first place. In Aarseth's view, a virtual object may be composed of a variety of virtual, real, and fictional aspects. While the resultant *ludo-realism* has similarities to Chalmers' later virtual realism, it diverges in its underlying attitude about what fictions are and how they are involved in virtual experiences.

In contrast, Mark Silcox takes up a partial defense of fictionalism about virtual objects. In Silcox's interpretation, part of the attraction of realism is motivated by the implausibility that we engage in the kind of "pretend" or "pretense" that would be required of fictionalism. We ought to reject fictionalism about virtual objects, the argument goes, when we realize that sophisticated users of VR do not treat virtual objects as *illusions*, as it would seem the fictionalist predicts. In response, Silcox argues that XR users take on a bimodal attitude, extended through time, toward virtual environments when they put on a headset. This attitude involves committing to undermine one's normal sensory experience (by means of the headset) while also fully acknowledging that one will continue to engage with the virtual as if this change has not occurred.

In his chapter, Grant Tavinor laments the extent to which metaphysical disputes between realists and fictionalists have dominated the conversation about virtual objects. Rather than engaging in speculative metaphysics, Tavinor encourages us to consider the underlying technologies and the way that their displays represent entities to users. Drawing on aesthetic theory and the history of art, Tavinor defends an alternative to metaphysical frameworks for thinking about VR environments. According to *VR picturalism*, VR experiences represent or depict their subject matter and so relate their users to the depicted subjects, in much the same way that a painting depicts its subject to an art viewer. Importantly, such depictions do *not* reproduce that which is depicted, either fictionally or otherwise, in Tavinor's view. As such, VR picturalism hopes to largely bypass the historic debate between realists and fictionalists that Aarseth and Silcox engage with more directly.

While the first three papers focus on how we should understand the objects and events within XR experiences, the final paper of this section explores the status of the *user* in XR. The view that the mind can extend beyond the physical organism of the human body has received increasing attention and support since Clark and Chalmers' (1998) defense of extended cognition. In their paper, Thomas Parsons and Joseph Neisser argue that the *self* may extend beyond the physical human body as well. Drawing on neuroscientific evidence and philosophical argumentation, Parsons and Neisser suggest that some sensorimotor extensions of the human body, including XR, could plausibly count as extending the boundaries of the self beyond consciousness and cognition. This places their view alongside other challenges to traditional body-based boundaries of the self (Gallagher 2013; Hayles 2017; Seth 2019). If Parsons and Neisser are correct, then entering an XR is not just a useful way of exploring previously neglected aspects of one's self. It could be an opportunity to *expand* the self across space and worlds. The possibility of an expanded self puts pressure on traditional notions of identity and ethics. As such, the paper serves as a segue into the topics discussed by authors in the second and third sections of this collection, as the extended self-view has interesting intersections with questions concerning ethics in XR and the use of XR for experimental research.

"Is There an Ethics for Extended Reality?"

The second section turns more directly to the ethical questions raised by Parsons and Neisser. To what extent can ethical principles and frameworks already deployed in non-virtual settings be applied in XR? There's a *prima facie* case to be made that little to no change is necessary. Call this the *simple answer*. If I promise to meet you at a certain time and place and fail to do so, I have plausibly failed in my obligations to you *independent* of whether the promise and proposed meeting occurs in a virtual or non-virtual space. It might be thought, then, that the process of determining ethics for XR mostly involves fitting pre-existing frameworks to virtual contexts.

This view falls short in at least three respects. First, it focuses primarily on *social* XR experiences, where there is interaction between agents that are already plausibly members of a shared moral community. It does not consider the possibility of ethics in solo experiences, including interactions with artificially intelligent bots and non-player characters (NPCs). Nevertheless, we might still wonder what human ethics and flourishing look like in such interactions. Second, it fails to consider how XR can be fundamentally transformative for XR users. For example, if Parsons and Neisser are right that the self can be extended across XR systems, then sabotaging a fellow user's virtual avatar may be more akin to assault than to property damage. Third, this argument focuses on ethical frameworks for in-experience user interactions. But the ethical concerns raised by XR extend to the developers of XR experiences, the companies and corporations that maintain XR hardware and software, and the variety of people (e.g., researchers, educators, politicians, etc.) that would use XR technologies in the pursuit of some further goal (Madary & Metzinger 2016). The papers in this section thus complicate and resist the simple answer.

Michael Madary frames his discussion of applied ethical issues in XR within the broader context of media more generally. Like other media,

XR has the power to reveal new ways to conceptualize ourselves and our experiences. But unlike previous media, XR more directly manipulates the primary medium: the body that mediates all experiences. As wearable technologies that can generate experiences of being in a different body and environment, XR can more directly shape our concepts of self and our relations to our surroundings. This grants the developers and maintainers of XR experiences and technologies great power over potential users. The ethical concerns Madary explores, then, are grounded in XR's status as a media technology and its related metaphysical questions. Without great care, Madary envisions the possibility of a world where users' self-conceptions and outlooks are curated by the goals set for them by corporations that control XR technologies.

It is precisely these questions of who controls embodied experiences and how they should be regulated that motivate the chapter from Erick Ramirez, Shelby Jennett, Dorian Clay, and Mohit Gandhi. XR software, from games that allow you to create custom virtual avatars to applications where you create virtual simulations of what you would look like in L'Oréal's newest makeup, allows you to virtually manipulate how you present yourself to others in technologically mediated spaces. In these ways, XR affords a kind of virtual bodily modification. However, standing ethical principles for understanding bodily modification are incomplete when applied to virtual contexts. As just one example, a user who finds LGBTQ+ relationships offensive could effectively erase queer users from appearing within their XR. Furthermore, artificially intelligent bots are possible that replicate the appearance and behavior of individual users. Thus, while XR may provide empowering opportunities for individuals to present themselves as they wish in virtual worlds, bad actors and bots may also disempower them. The chapter weighs these difficulties and others, as well as the difficulties of using XR technologies to try to address them.

The final chapter of this section focuses more exclusively on interactions with bots by considering the ethics of virtual actions in solo XR experiences. Many are tempted to think that virtual actions performed in self-contained (non-social) XR contexts are amoral, perhaps motivated by the *simple answer* described above. After all, the thinking goes, no one is harmed. Andrew Kissel argues that, at least in some cases, it may be appropriate to make moral assessments of individuals for their virtual actions, even in solo XR. Crucially, such assessments are only appropriate when the action can be properly attributed to the user, and not merely to a role that the user adopts temporarily for the XR experience. Determining the difference, according to Kissel, depends on our understanding of the self and identity. Adopting a narrative theory of identity, Kissel argues that virtual actions can, in some cases, partially constitute one's narrative self. An action is properly attributable to a person when it can be made sense of in terms of the ongoing narrative that person constructs for their life. When this requirement is met, a user's virtual actions are expressions of themselves in a way that can ground moral assessments. Resonating with themes in other chapters, the picture suggests that experiences in XR involve the extension and constitution of the self in virtual contexts. Furthermore, it suggests ways that previous frameworks can be extended but must also be modified in order to account for this new ethical frontier.

What Can We Do with Extended Realities?

The third section takes a more empirical turn. It examines the psychology of simulated experience and moral cognition in XR experiences, with a particular eye to the way these experiences have (and could) be used for research and education. XR technologies create experiences of presence, as defined above, which has been seen as key to many of its successful applications, especially training and educational simulators. The impact of presence has also made XR an exciting tool for psychological research. As explored in earlier papers in this collection, some researchers are using XR experiences to balance concerns about control and ecological validity (i.e., will results in the lab hold up in actual environment applications?) with practical concerns about experimentation (i.e., could this experiment actually be run?)

In her chapter, Kathryn Francis presents research on the judgment–action discrepancy in moral psychology. Is what people *judge* to be morally acceptable in a cool moment of reflection consistent with the *actions* they perform in moral situations? To address this question, Francis presents the life and death "trolley problem" moral dilemma in VR. In the dilemma, participants must choose whether to take an action that preserves five lives at the cost of one life or to refrain from acting in order to preserve one life at the cost of five. Obviously, putting actual lives in danger to study their moral responses is out of the question. So historically, the dilemma has been presented via text. But this method gauges moral *judgments*, not necessarily moral *actions*. The immersive features of XR provide a more direct way to investigate moral actions. Beyond moral psychology, the affordances of XR greatly open the possibilities for researchers. However, these new paradigms are not without difficulties, as Francis acknowledges. The widespread adoption of moral psychology must deal with ethical concerns for the psychological wellbeing of the participants, as well as further research into the increased ecological validity claims for XR paradigms.

Key to addressing these ethical concerns is greater investigation into the psychological effects of XR, a project undertaken in the chapter by Eugy Han and Jeremy Bailenson. Once again tying together ethics, metaphysics, and psychology, their approach focuses on both the behavioral changes of people in XR as well as the XR environments that impact them. For example, they describe research on the "Proteus effect", wherein user behavior changes to fit their digital avatars, independent of how other people view the avatar. In short, what you look like in virtual spaces ends up affecting *how* you behave. This and other research Han and Bailenson explore reraise previous ethical questions about who controls bodily presentation in XR while also raising new questions about the impact of XR experiences on users going forward. As communities in XR develop further, it is increasingly clear that social experiences in XR have many similarities but also numerous differences from social experiences outside of XR. Keeping in mind Madary's earlier discussion of XR as a form of controlled (and monetized) media, knowledge of the psychological impact of XR and XR communities can be used to shape and nudge users themselves. Whether this will be a good or a bad thing remains to be seen.

In the final chapter of the section, Jon Rueda explores ways of using XR to improve morality. Drawing on moral enhancement literature, Rueda argues that the unique affordances of XR could be used to enhance empathy, reduce implicit bias, and improve pro-environmental behavior, among other potential moral benefits. Rueda warns that such projects should not be undertaken lightly, as there are numerous potential pitfalls in advancing such programs. Not the least of which are concerns about what counts as "improved" moral behavior. Nevertheless, and echoing themes from the other two chapters in this section, the breadth of experiences XR allows for, and the ways that XR can affect us, suggest we should start thinking about what kind of life we want to live, both inside and outside of XR experiences, now. The future is already here.

The collection closes by explicitly turning to questions in the near future of XR, motivated by the articles in this collection. It is an admittedly cloudy future, with major XR projects halting (e.g., Mojo Vision's AR contact lenses being abandoned in January of 2023) at the same time that new contenders enter the field (e.g., Apple's announcement of their first XR headset in June 2023). One of the principles that have driven this collection is a concern for the metaphysical, ethical, and psychological questions that are arising from XR technologies as they already are, here and now, rather than just thinking about far-flung sci-fi futures. This approach runs the risk of making the questions under consideration appear relevant only with respect to the state and form of the technology as it is now. On the contrary, we suggest that each article in this collection charts exciting new possibilities for further research and discussion on XR and its social, political, and transformative potential that continue to develop along with the technologies that ground them.

References

Chalmers, D. (2022). *Reality+: Virtual Worlds and the Problems of Philosophy.* UK: Penguin.

Clark, A., & Chalmers, D. (1998). The extended mind. *Analysis, 58*(1), 7–19.

Cummings, J. J., & Bailenson, J. N. (2016). How immersive is enough? A meta-analysis of the effect of immersive technology on user presence. *Media Psychology, 19*(2), 272–309.

Gallagher, S. (2013). A pattern theory of self. *Frontiers in Human Neuroscience, 7,* 443.

Hayles, N. K. (2017). *Unthought: The Power of the Cognitive Nonconscious.* Chicago University Press.

Madary, M., & Metzinger, T. K. (2016). Real virtuality: A code of ethical conduct. Recommendations for good scientific practice and the consumers of VR-technology. *Frontiers in Robotics and AI, 3.* https://www.frontiersin.org/articles/10.3389/frobt.2016.00003/full%20, https://doi.org/10.3389/frobt.2016.00003

Nilsson, N. C., Nordahl, R., & Serafin, S. (2016). Immersion revisited: A review of existing definitions of immersion and their relation to different theories of presence. *Human Technology, 12*(2), 108–134.

Pokémon, G. O. (Iphone Version). [Video game]. (2016). Niantic.

Seth, Anil K. (2019). Being a beast machine: The origins of selfhood in control-oriented interoceptive inference. In Columbo, Irvine & Stapleton (eds.), *Andy Clark and His Critics* (pp. 238–254). Oxford University Press.

Vi, S., da Silva, T. S., & Maurer, F. (2019). User experience guidelines for designing hmd extended reality applications. In *Human-Computer Interaction–INTERACT 2019: 17th IFIP TC 13 International Conference, Paphos, Cyprus, September 2–6, 2019, Proceedings, Part IV 17* (pp. 319–341). Springer International Publishing.

Weber, S., Weibel, D., & Mast, F. W. (2021). How to get there when you are there already? Defining presence in virtual reality and the importance of perceived realism. *Frontiers in Psychology, 12.* https://www.frontiersin.org/articles/10.3389/fpsyg.2021.628298/full%20, https://doi.org/10.3389/fpsyg.2021.628298

Part 1

What Is Extended Reality?

1 "Back to Reality"

The Case Against Ludo-Fictionalism

Espen Aarseth

Introduction

Notions of ludic immersion and games as fictions have dominated the philosophical and aesthetic discussions about (especially computer and video) games from the start. The academic field of game studies, while institutionally quite recent (see Aarseth 2001), has roots back to Johan Huizinga's classic study of games as a foundation of culture, *Homo Ludens* (1938), in which he posits the special otherworldly quality of playing games as a zone of experience divorced from ordinary reality, an immersive "magic circle." Before game studies coalesced as a field, a large amount of highly relevant work was done in virtual reality (VR) research, both technical, sociological, and philosophical, especially in the '80s and '90s. There seems to be a watershed moment around the fall of the first wave of VR around 1995 and the rise of 3D games and, a little later, game studies in the late '90s. Many of the same topics and research problems were carried over or rediscovered by games researchers not familiar with the VR literature, just like they are again today, by philosophers and VR researchers not familiar with the early game studies literature.

The central notion of "immersion" has been questioned (Salen & Zimmerman 2004) and strongly criticized (Calleja 2011), but the core issue of this chapter, the problematic notion that events in VR and games are fictional (that ludic or virtual representations point to fictional content) is still current. However, it soon became obvious in game studies that (especially) multiplayer games represented a new kind of social reality in which the fictional aspects had to share the arena with social behaviors that had little to do with fiction, e.g., Castronova's observation (2001) that *EverQuest* money had real value. Similarly, games like *Minecraft* can be used to build complex machineries – players have implemented computer games like Pong and even in-game programmed a simulated Gameboy handheld console running a version of *Pokemon FireRed*[1] – which, due to their technical reality, cannot be classified as fictional. Also, games can have

DOI: 10.4324/9781003359494-3

a documentary frame of reference (e.g., *The Oregon Trail*, 1974), which leads to the observation that it is necessary to distinguish between games (and VR) with a fictional and a documentary referentiality (See Fullerton 2008 for an excellent discussion of ludo-documentary representation). Indeed, much of the later claims of game fictionality seem to conflate the notions of fiction and representation, following Juul (2005). In *Half-Real* (2005: 139), Juul discusses how experienced Quake3 Arena players would turn down the graphics to get better performance:

> With sustained playing of the same game, the player may become less interested in the representational/fictional layer of the game and more focused on the rules of the game.
>
> (Juul 2005)

Here, it seems that Juul (and many others after him) is conflating fiction and representation since fiction must always be represented, not representational. Or, to follow de Saussure, a fiction is a signified, not a signifier. A signifier used for fiction is not in itself a fiction and can also be used for non-fiction. However, game signifiers do almost always look like typical signifiers used by fiction (e.g., cartoonish or hypermasculine), so the conflation is easily explained. But if a historian of popular culture discusses, say, Mickey Mouse, and uses the name and also the image, for instance in the sentence "Mickey Mouse was Walt Disney's first successful cartoon character", the signifier MM references the culture-industrial icon and its historical status, not the fictive figure and his escapades.

Furthermore, if the frame of reference of a game or VR installation can be either fictional or documentary, then their representations cannot be fictional *per se*. For many games researchers in the 2000s, the highly influential games-of-make-believe fiction theory of Kendall Walton (1990) led them to assume that games were a kind of make-believe, and therefore fictional, but all games are of course not games of make-believe. Walton can be suspected of using "game" metaphorically, or at least in a non-standard way, since his subjects would not, typically, argue that they were playing a game when they were engaging in the practices he describes, such as looking at a painting. Thence, this paper will argue that Walton has been critically misunderstood and that his theory does not apply to games in general, or even to games in virtual worlds. These games quite explicitly focalize the objects that players can operate or act on, and so they are not, as the Waltonista claim, Waltonian props, but rather very concrete virtual tools with very concrete functions and affordances (a prop is a non-functional object whose only necessary property is a visual(/auditory) resemblance to the real one, in which case I should be able to use a shotgun as a rifle when I play *CounterStrike,* or even a virtual stick, but I cannot).

With the recent entry of analytical philosophers into game studies (e.g., Tavinor 2009; Meskin & Robson 2012, Tavinor this volume), ludic fictionalism has entered yet another, and perhaps more insular phase, where the philosophers mostly focus on the works of other philosophers and are primarily trying to establish the concept of fiction in a new empirical field, while ignoring alternative perspectives and terms like virtual and simulation. In this paper, expanding on Aarseth (2007, 2011, 2014), I will address their claims about games and fictionality, and show how the ludo-fictional hypothesis fails to explain problematic or even obvious examples like the above.

Rather than relying on fictional content to involve the player in a game of make-believe, games in virtual worlds engulf their players by offering real, but non-physical alternatives as instruments of play and exploration. There is nothing primarily fictional about a virtual gun in a computer game; its empirical properties allow the same kinds of manipulation that a computer user has available when, say, they are typing a scholarly paper in preparation for a workshop and filing it in a document folder. During the writing, black marks on a white background form exactly as if a classic Remington typewriter had been used instead. Neither the gun nor the typed paper is physically tangible, but both exist in ways that a fictional object would not, and they are not any kind of prop but instead represent themselves the same way a physical gun or material manuscript would. In insisting on the fictionality of virtual objects, the ludo-fictional thesis fails to grasp the most crucial ontological turn of our time: from material to informational reality. So much of what we do on a daily basis, and especially some of our most important activities, are mediated through cybernetically enhanced perceptual technologies, and yet we would never think of them as unreal. (However, we sometimes start out with this very suspicion: the adjective "phony" was derived from the early telephone experiences.) It is the empirical reality of games that makes us lose ourselves in them, the sensation of actually being there, doing that.

What Is Fiction? Computer Games and Fiction: A Brief Review

"Fiction" is, or used to be, a vernacular, non-scientific term and concept, used in public discourse for centuries. For a long time in academia, it was synonymous with literature, as in "literary fiction", denoting a genre that was not lyrical poetry and not drama, nor essay, or historiography, etc. The first writings on games as fiction appeared in the eighties, with pioneers like Mary Ann Buckles, who wrote the first PhD dissertation about a computer game in 1985. The earliest use of the phrase "Interactive fiction" might be found in Bob Liddil's article in Byte in 1981 (Liddil 1981). These early examples did not typically reflect on the notion of fiction but used it rather uncritically and synonymously with Literature. Later, I observed that simulative games like *Adventure* "are somewhere in between reality and fictionality: they are not obliged to represent reality, but they do have an empirical logic of their own, and therefore they should not be called fictions" (Aarseth 1994: 79). I followed this up in a short article about the hermeneutical challenges of virtuality (Aarseth 2001) and expanded it into a theory of the nature of game objects (Aarseth 2007) in which I posited that game objects are often ontological conglomerates: The same game object can have fictional, virtual, and real aspects. This position of ludo-realism (see also Aarseth 2014) is virtually identical to Chalmers' of *virtual realism* in "The Virtual and the real" (2017) but, as we shall see, contains interesting nuances and departures. Chalmers slightly misrepresents my position (Aarseth 2007: 39) for one that holds that virtual worlds are not real but instead related to dream worlds and mirror worlds (2017: 6f5); but it is more complicated than that, and in my view, all of these worlds are different from each other, and so, potentially, are the individual game worlds. The position promoted in Aarseth (2007, 2011, 2014) is that game worlds are indeed real at least in part, but that both ludic and virtual worlds also can contain fictional material, just like the physical world can (e.g., a trompe-l'œil). As mentioned, games may contain fictional, virtual, and real elements. A virtual element is one where the physical aspect is simulated, but non-physical aspects could still be real; in other words, the concept of virtuality only seems meaningful when linked to fauxphysicality and is not a useful descriptor of non-physical phenomena in game worlds. A conversation between two players in a game world is not a virtual conversation.

At some point in the '80s, the question "What is fiction?" became an interesting problem for analytical philosophy, and then the race was on to define fiction, and to turn it into a contested theoretical concept that it hadn't been before. (Up until then, literary theory had used the word rather complacently, in its vernacular sense.) This sort of definitional competition happens relatively often, and typically with similar problems. One example is the semi-scientific category "planet", which may or may not include Pluto. Pluto's planethood has been decided by a few rounds of voting among astronomers in 2006 (mostly voting no) and lay people in 2014 (overwhelmingly voting yes). Curiously, the "planet" category does include both Pluto-like hard objects (Mars, Earth) and Pluto/Earth-unlike, gaseous ones (Jupiter, Neptune), but for many astronomers, not Pluto itself. Turning a folk word into a theory word may or may not be a good idea (because this sort of semantic hijacking can piss ordinary people off), but the test is in the result: does the definition serve analysis and become a fruitful perspective, is it too costly to maintain, or does no consensual definition emerge at all? It appears that fiction can be defined in numerous ways, of which there are at least three reasonable positions, which parallels the familiar sender-message-receiver triptych of (textual) communication or, in other words, follows the Aristotelian distribution between Ethos, Logos, and Pathos

1 Fiction is best described through sender's intent (G. Currie 1990);
2 Fiction is a referent (text) without a (real-world) reference (D. Cohn 2000); and
3 Fiction is best understood through receptive practices (S. Friend 2008).

Enter Kendall Walton (1990) – fiction is a prop in a game of make-believe. Walton holds that a work of fiction is a prop that prescribes imaginings. This is a very wide category; drawings and even photographs are included. An extremely useful critique of Walton's position has been made by Stacie Friend (2008).[2] Friend points out that "Walt-fictions," as she calls them, are not standard fictions, but so much more, and a work or textual phenomenon can be a Walt-fiction without being a fiction (per other, more moderate definitions).

To a ludo-realist, Walton's otherwise impressively radical theory is rather annoying, since it derails the "are-games-fictions?" debate: Games could be (A) Walt-fictions without being (B) fictions. A number of otherwise highly capable game scholars have been conceptually poisoned by Walton's dangerous idea, e.g., Walker [Rettberg] (2003) and Van Looy (2006), to unfairly single out a few. However, even Walton himself is skeptical of (A); in a very revealing interview with Chris Bateman (2010), Walton suggests that many games are not Walt-fictions:

Bateman: What about chess… do you think a chess player plays a game of make-believe with their pieces?
Walton: I doubt that chess players ever engage in make-believe. The game would be no different if pieces were called "Piece #1", Piece #2" etc., rather than "King," "Queen," "Bishop," and so forth.
Bateman: But the pieces are clearly at some level representative of Kings, Queens, Bishops and Knights, or at least they once were…
Walton: Those names do indicate a kind of make-believe players could engage in, and perhaps did in the ancient past. It is likely to be what I call a *prop oriented* game, however, rather than a content oriented one, i.e. a game in which our interest is in the props rather than the fictional world that the props generate. [...] The make believe in which novels and stories are props are largely content oriented, whereas many metaphors involve prop oriented make-believe. We might think of the names of chess pieces as dead metaphors – recalling perhaps a previous activity of prop oriented make-believe, but one we no longer engage in.

Here Walton raises an interesting alternative: whether game tokens should instead be seen as dead metaphors. Metaphor theory, as Sebastian Möring (2013) has shown, can indeed be very useful in explaining aspects of game representation, but it would be too expansive to go off in that direction in this short paper. Let's pretend to ignore Walton's skepticism for now, and instead take his followers seriously. In brief, this leads me to the following four problems:

1   Are games Walt-fictions?
2   Are game objects representations?
3   Is fiction theory a suitable or sufficient ontology for games?
4   Are game objects "props"?

1   Games in general are not Walt-fictions. Some games are, those that can correctly be termed role-playing games (e.g., tabletop campaigns, LARPs), games that rely on the imaginations of players to function at all. For the rest, they can be played with interchangeable representational layers, as pointed out in Aarseth (1997: 40), and so they do not depend on any specific set of representational components. What complicates this argument is that video "games" are often not mere games; they are entertainment software packages that contain fictional (or documentary) discursive material in addition to the game components. These non-game aspects of games can be said to be both fictional and Walt-fictional (e.g., the more than seven hours of cut scenes in *Metal Gear Solid 4*), but since the game parts of these complex packages can be reskinned or even kept non-representational ("abstract"), it is not the case that games in general (GIGs) are Walt-fictions.

2   Game objects are not representations. Meskin and Robson (2012) hold that "we take it as incontrovertible that videogames belong to the class of representations" and "[e]ven Tetris plausibly involves Waltonian representation". Let's take the first claim first. [Video-]Games, like all systems which make use of communicational acts for their operation, necessarily must *use* some form of representation. This, in my view, does not entail that they themselves *are* representations in a fundamental sense. A more moderate position would be that some game software packages are representations in a fundamental or reasonable sense, but that the rest merely use representations and that the representational aspect is arbitrary (see Aarseth 2014 for *The Suicide bomber/Dean for Iowa* example of two games with crucially different representations but identical gameplay). So-called reskinning (changing the representational layer of the game) typically does not change the ludic aspect, but it may change the interpretational potential. Most games, and even videogames, however, do not, as long as they are not game-story hybrids

or ludo-narratives, prescribe "interpretation" to any substantial degree; they merely demand operational response. As for Tetris, it can be role-played, yes. Any game or indeed any system or act at all, say, driving a bus, can be used as a prop for roleplay, but then this addition of a role-playing layer turns the activity into a different game or system. Tetris does not prescribe roleplay, only goal play, and, therefore, it does not depend on Waltonian representation. Games, like humans, ATMs, and vacuum cleaners, employ representation as necessary but insufficient aspects of their operation. They should not, therefore, be reduced to this one dimension; they are complex objects, conglomerates of representation and functionality, and by focusing on only one dimension of these conglomerates, it is difficult to make accurate claims about their nature. In addition, the ontology of games and video games is still poorly understood, and they are often conflated into one homogenous medium, like film or the print novel, which they are not. Even the fairly wide notion of VR is much more concrete and narrower than (video) games.

3 A fiction theory (any of them) is not a suitable or sufficient ontology for games. The concept of fiction has been shaped by centuries of studies of literary prose, and, therefore, it seems to make less and less sense proportional to the distance from its originatory genre. Even film and drama have quite different conditions of representation to literature, to the extent that it is not common to call them fictions in everyday language use. At first, there were little or no reflections accompanying the application of "fiction" to games, and this is still largely the case in game studies (and analytical game studies) today. There are (too) many phenomenological, empirical, linguistic, economic, ethical, and material arguments against simply calling games fictions, or even Walt-fictions. Perhaps the ball should be in the other court: what is gained and (to quote Joni Mitchell), what is lost, or, what can one imagine will be lost by placing such a complex and superdiverse family of phenomena in such a narrow category? *Cui bono?*

4 Game objects are not props. A prop (orig.: a property owned by a stage actor) is a stand-in for a fictional object: this stick is a sword, that stub is a bear, this stage is a desert. There is a purely representational, non-functional relationship between the prop and the fictional reference. A usable game object, on the other hand, is a functional object that will directly support the player's operational play, and not (merely) pre-scribe imaginings. So, a functional game object (car, gun, landscape) is not a prop, but part of a primary affordance that makes a strategic difference to the player in the game world. It does not represent; *it is*, in the most basic sense, *useful*. Tavinor (2009) attempts to get away from this problem by using the phrase "robust prop." However, this indicates that he recognizes crucial differences between game objects and normal

props but is still unwilling to depart from the Waltonian perspective and nomenclature. Also, to export the notion of prop from the theatre and into the domain of computer games and VR is a proposition that needs to be justified by arguments; it does not follow naturally or logically that the relation "prop" between a physical theatre object and its fictional reference is the same as the relation between an object in a simulation and the object it simulates.

What Is a Digital/Virtual/Game Object?

David Chalmers' essay on the real and the virtual (2017) presents a strong case for virtual realism, a position that is very close to what I elsewhere have called ludo-realism (Aarseth 2014): game objects and game events are real, or at least closer to reality than to fiction. This position was minted during the height of the first VR wave: "Simulations are somewhere in between reality and fictionality: they are not obliged to represent reality, but they do have an empirical logic of their own" (Aarseth 1994: 79). Where Chalmers and I may diverge is on a crucial aspect for us both: the nature of virtual/digital/game objects. Here, it is vital to realize that these may not have the same reference; in my view, an object can be virtual without being digital, and vice versa. For Chalmers, the focus is on the ontology of technological VR, that is, computer-generated worlds and objects in those worlds. His paradigmatic virtual world is not a game but *Second Life*, accessed via a VR interface. In contrast, what I call *games in virtual environments* include any game that simulates some physical world, whether the simulation is computerized, mind-based, tabletop/board-based, or otherwise mechanical. Also, for me, not all game objects in a digital game are virtual, some, like a labyrinth, are purely real (conceptual/intentional), and others, like a game character, are fictional or documentary (but when documentary then also, typically, counterfactual). A non-player humanoid (or animaloid/monster) is typically composite: some aspects are real (the operational), some are virtual (the simulated physiology), and some can be fictional or documentary (the personality). I agree with Chalmers that VR worlds like *Second Life* (SL) can be a simpler basis for the discussion of the real vs the virtual since the question of fiction does not get so much in the way. Therefore, I will handicap my perspective to this sort of phenomenon from here on.

Counter to Chalmers' position, it would seem to me that "digital objects" in SL is not one category, but many, some virtual, some not, so digital and virtual should not be used as synonyms. For instance, there are many texts ("books") and videos to be found in SL. In such cases, the texts – strings of letters – found in SL books are not virtual (although their "physical" aspects may be), but they would still be digital. An image or a video on a wall in SL would not be a virtual image or video, it would simply be an image or video, or a digital image or digital video. A labyrinth in SL would not be virtual in my sense (see Aarseth 2007), but presumably in Chalmers'. More fundamentally, the simulated materiality of SL objects can make them behave in many different ways: some are impenetrable; others can be walked through as though they have no materiality. This means that they are not, on a fundamental level (embodiment), the same type of object. Chalmers (2017) also seems to adopt this position when he discusses how some digital objects are real and holds that "the digital object corresponding to a virtual library is not just a virtual library: it is really a library" [14]. Perhaps by the same logic, are some digital objects fictional? A door that has no functional properties, but is just a door image tacked to a virtual wall, is not a virtual door, but a fictional one, or perhaps a tromp-l'oeil in VR. The ontological status of doors in VR is explored further in Aarseth (2007).

One could also imagine personalized filters which would allow for individual experiences; one person sees and hears multiple other avatars or objects in the same VR room; another perceives only one, or none, just like blocking in social media. Can our notion of the real cope with this? If SL-like, more advanced technologies become the social norm (rather than today's ludic VR), might not "post-reality" be a more fit description than reality, since we would not be bound by the same basic perceptions? Reality expands as new modes of perception become available, through science and technology, but it still presumes and is checked by social consensus. However, as we have seen lately, social technology also allows for diverging "realities" where the flow of information can be tailored to individuals, with or without their knowledge.

If we have to choose between "fictional" and "real" to describe SL-like phenomena, then I agree that "real" is the best and only reasonable choice. But why are we limiting ourselves to that choice? Just because "fiction" is a poor conceptual fit, it does not follow that "real" has all the analytical power needed, especially if we are experiencing genuinely new material constellations of human construction, such as the current revolution in artificial intelligence.

Notes

1 https://www.youtube.com/watch?v=paoEeRG-j8U
2 I am indebted to Meskin and Robson (2012) for making me aware of Friend's essay.

References

Aarseth, E. (1994). Nonlinearity and literary theory. In Landow (ed.), *Hyper/Text/ Theory* (pp. 761–780). Johns Hopkins University Press.

Aarseth, E. (1997). *Cybertext*. JHU Press.

Aarseth, E. (2001). Virtual worlds, real knowledge: Towards a hermeneutics of virtuality. *The European Review*, 9 (2): 227–232.

Aarseth, E. (2007). Doors and perception: Fiction vs. simulation in games. *Intermédialités*, 9: 35–44.

Aarseth, E. (2011). Define real, moron! *Digarec Series*, 6, 50–69.

Aarseth, E. (2014). Ontology. In M. J. p. Wolf & B. Perron, (eds.), *Routledge Companion To Video Game Studies* (pp. 484–492). Routledge.

Bateman, C. (2010). Walton on make-believe.Retrieved from http://onlyagame.typepad.com/only_a_game/2010/06/walton-on-makebelieve.html

Buckles, M. A. (1985). *Interactive Fiction: The Storygame 'Adventure.'* Dissertation. University of California, San Diego.

Calleja, G. (2011). *In-Game*. MIT Press.

Castronova, E. (2001). Virtual worlds: A first-hand account of market and society on the cyberian frontier. *The Gruter Institute Working Papers on Law, Economics, and Evolutionary Biology*, 2 (1): 1.

Chalmers, D. J. (2017). The virtual and the real. *Disputatio*, 9 (46): 309–352.

Cohn, D. (2000). *The Distinction of Fiction*. JHU Press.

Currie, G. (1990). *The Nature of Fiction*. Cambridge University Press.

Fullerton, T. (2008). Documentary games: Putting the player in the path of history. In Z. Whalen & L. Taylor (eds.), *Playing with the Past: History and Nostalgia in Videogames* (pp. 215–238). Nashville: Vanderbilt UP.

Juul, J. (2005). *Half-Real*. MIT Press.

Liddil, B. (1981). Interactive fiction: Six micro stories. *Byte*, 6 (9): 436.

Meskin, A., & Robson, J. (2012). Fiction and fictional worlds in videogames. In *The philosophy of Computer Games* (pp. 201–217). Dordrecht: Springer Netherlands.

Möring, S. (2013). *Games and Metaphor: A Critical Analysis of the Metaphor Discourse in Game Studies*. Ph.D. Thesis, Copenhagen: IT University of Copenhagen. URL: http://bit.ly/1Gtcg4k.

Salen, K. & Zimmerman, E. (2004). *Rules of Play: Game Design Fundamentals*. MIT press.

Tavinor, G. (2009). *The Art of Video Games*. Wiley-Blackwell

Van Looy, J. (2006). *The Promise of Perfection: A Cultural Perspective on the Shaping of Computer Simulation and Games*. Dissertation, U Leuwen.

Walker[Rettberg], J. (2003). *Fiction and Interaction: How Clicking a Mouse Can Make You Part of a Fictional World*. Dissertation, U Bergen.

Walton, K. L. (1990). *Mimesis as Make-Believe: On the Foundations of the Representational Arts*. Cambridge, MA: Harvard University Press

2 Fictionalism and Virtual Objects

Mark Silcox

In most people's everyday mental and discursive lives, beliefs, and assertions about fictional entities blend seamlessly together with ordinary thoughts about the empirical world. Inferences about real-life mysteries can involve speculations about what Sherlock Holmes would have thought, real causal links between historical epochs are often understood as passing through intermediate, made-up events, and practical wisdom about how to live is frequently gained through contemplating the actions performed by beings of pure fantasy. Philosophical commonsense would seem to dictate that, in spite of all this, the most accurate models of human cognition will find a way to sequester whatever it is about us that allows for the construction of mere fictions from the mental faculties that connect our most empirically well-founded beliefs to the "real" world. But what if some of the types of objects that philosophers have traditionally thought of as absolutely fundamental constituents of reality – numbers, say, or word meanings, or Platonic forms – might also turn out to be best regarded as fictional? What sort of an explanation would be sufficient to reconcile the relative contingency of the fictional imagination with the absolute necessity of some of its proper objects? It is this general concern that underlies the contemporary discussion about fictionalism in metaphysics.

It seems to me that investigating the singular style of thinking that takes place when human beings interact with VR technologies can provide some insight into this heavily contested debate. In this paper, I want to investigate the possibility that the special types of objects that populate virtual environments are also themselves best thought of as fictions. My goal will not be to defend this view unequivocally, but rather to present it as a sort of case study for illuminating the plausibility of metaphysical fictionalism more generally.

My stalking horse for most of the discussion will be an influential general argument against fictionalism originally made by Jason Stanley in his 2001 paper "Hermeneutic Fictionalism." An exploration of different strategies for responding to Stanley's argument will lead into a detailed

DOI: 10.4324/9781003359494-4

examination of the defense provided by David Chalmers of metaphysical realism about virtual objects in his (2017) essay "The Virtual and the Real," as well as his more recent book *Reality+* (2022). In the course of pointing out some subtle ambiguities in Chalmers' characterization of the VR user's cognitive states, I shall try to illuminate an important difference between two ways in which an object of human thought can merit the label "fictional" that does not seem to me to have been adequately distinguished in the literature. This will open the door for a fairly decisive response to Stanley's criticism that at least significantly strengthens the case for fictionalism about virtual objects. I shall conclude with a few suggestions about the implications of this proposal for broader issues in metaphysics.

The Fictionalizing Unconscious

Like most other well-known anti-realist and irrealist positions in metaphysics, fictionalism proposes an account of our speech and thought about a class of putative entities that is designed to explain both (i) why reference to and descriptions of the relevant entities might make them *seem* real and (ii) the attitudes and motivations of speakers and thinkers who perpetuate these semantic illusions. Explanatory strategies of this general flavor have been commonplace in philosophy since long before contemporary versions of fictionalism came upon the scene, perhaps most famously from defenders of nominalism about universals[1] and advocates of expressivism in ethics.[2]

Contemporary fictionalists often talk about *make-believe, pretense,* or other acts of the imagination in their attempts to satisfy requirement ii. Thus when somebody says "Red is a warmer color than purple," they are only *pretending* to compare two discrete objects of perception; when someone says "Eating chicken is wrong," they are only *pretending* that acts of cannibalism possess a moral property that is as real as whoever is eating or being eaten. But it is sometimes difficult to tell whether such philosophers are using these psychological terms to classify just *any* sort of attitude on the part of speakers and thinkers that represents unreal objects as real, or whether they have much more psychologically specific phenomena in mind.[3] This ambiguity provokes a justified suspicion that it might turn out to be impossible to zero in upon exactly what sorts of mental states are in operation when we speak and think about fictional entities, not only from a detached scientific or philosophical perspective, but from the perspective of one who is allegedly engaged in the relevant types of speech or thought.

Stanley's critique of fictionalism is directed exclusively at what he calls the "hermeneutic" variety. He characterizes this view as a hypothesis concerning how the relevant discourse (e.g., about numbers, universals, epistemic states, moral virtues) "is in fact used." The hermeneutic fictionalist believes that speakers who use such vocabulary normally do not mean their claims to be taken as truth-evaluable at all, but only pretend to. Stanley contrasts this attitude with what he calls "revolutionary" fictionalism, which would instead consist of a proposed "reconstruction or revision" of the discourse in question.[4]

Stanley takes the hermeneutic fictionalist to task for supposing that anything as psychologically sophisticated as make-believe or "pretending" could be going on during our everyday thought and talk about the types of entities purported to be fictional:

> [I]n any case of interest, the hermeneutic fictionalist's position entails that whether or not someone is engaged in a pretense is inaccessible to that person...Now, pretense is unquestionably a psychological attitude one bears to a content; it is in the same family of attitudes as belief....If the hermeneutic fictionalist is correct, then *x* can bear the propositional attitude of pretense towards a proposition, without it being in principle accessible to *x* that *x* bears the propositional attitude of pretense towards that proposition. But this introduces a novel and quite drastic form of failure of first-person authority over one's own mental states.[5]

Given ordinary language users' apparently widespread felicity in communicating with one another about such matters as mathematical relations and moral judgements – and especially, perhaps, given the possibility of expertize upon such topics – it is *prima facie* at least a little bizarre to suggest that the norm in our discussions thereof is a widespread failure to be aware of our own semantic intentions.

The first thing worth mentioning in response to these observations is that most of us are actually familiar enough with many everyday situations in which individuals engage in pretense without full self-awareness, and within which, furthermore, acquiring access to this fact about their own mental states might be considerably effortful (or even psychologically painful) for them. Think for example of the way that fans of professional wrestling discuss the antics of their favorite pugilists, or the way that religiously observant non-fundamentalists sometimes talk about the ostensibly historical portions of the Bible. So the suggestion that, in at least some commonplace circumstances, the attitude of pretense might be *in principle* inaccessible to somebody who adopts it is at least a little less "novel" and "drastic" than Stanley makes it out to be.

Nonetheless, perhaps it's not an exaggeration to say that most types of pretense are *normally* performed self-consciously. And if this is true, then in circumstances where those who engage in specifically existential pretense need to have this very fact pointed out to them, the metaphysical fictionalist is subject to a special explanatory burden of accounting for why this is so. Certainly, in the sorts of cases that have attracted the most attention in the mainstream literature about fictionalism – claims about mathematical objects and *ante rem* universals,[6] and attributions of propositional attitudes such as knowledge or moral insight[7] – this burden seems especially challenging to discharge since activities such as performing everyday mathematical calculations or explaining intentional human activity don't feel very much like make-believe to ordinary participants.

There are at least a couple of features of human interactions with objects in virtual environments that might make it seem even more challenging for the VR fictionalist to respond to Stanley's phenomenological objection. First of all, one of the chief characteristics that have been regarded as definitive of VR for at least as long as philosophers have been writing about it is *immersivity*. This property of virtual environments has usually been characterized in terms of both the exclusion of extraneous stimuli and the user's sense of "presence" – the feeling, that is, of being affected directly and without mediation by one's virtual surroundings.[8] When somebody wearing an HTC Vive hears a real car screech by outside, she is likely to attend to it far less than she would without the headset on. But when she sees a virtual Frisbee speeding toward her, she ducks, when a virtual monster jumps out at her from behind a virtual sofa, she screams. These sorts of reactions surely often happen far too swiftly and spontaneously for the user to cultivate even the most fleeting intention to make-believe or pretend anything whatsoever. Secondly, VR objects clearly differ from many of the other types of entities philosophers have claimed to be fictions insofar as they have a much more discretely and easily identifiable basis in the physical world. The connections between numbers, Platonic forms, or instances of intrinsic moral goodness and the material realm are infamously difficult to specify. But whatever else we know about that Frisbee we see in the virtual world, it certainly stands in some pretty strong relation of metaphysical dependence to goings-on inside a relatively small, confined space in the non-virtual world – *viz.* the VR headset itself and whatever other devices are being used to run the relevant program.

In his defense of realism about virtual objects, Chalmers makes no direct appeal to the phenomenological objection to fictionalism. But the account he gives of the intentions and attitudes of VR users is meant to secure the plausibility of realism *via* an account of the mental states that distinguish a "sophisticated" user of the relevant technologies. I shall examine this account closely in the next section to tease out some of its phenomenological and metaphysical implications, and to see how these compare to the type of story the fictionalist might tell.

How to Be a VR Sophisticate

According to Chalmers the central distinguishing feature that differentiates virtual mountains, tables, and calculators from the ordinary objects we interact with every day is that the former are "digital objects" grounded in "data structures." These structures are themselves grounded in the physical world that grounds non-virtual objects more directly.[9] But according to the general program of epistemic structuralism Chalmers subscribes to, *all* real and virtual objects are *also* metaphysically grounded in our perceptions.[10] It is the account he gives of the specific features of how we apprehend the objects that inhabit VR, and his attempts to contrast it with the view (which he attributes to fictionalists) that such perceptions are *illusory*, that does most of the work in Chalmers' defense of "virtual realism."

This insistence upon a connection between fictionalism and an illusion theory of perception in VR is itself quite tendentious. An even more problematic assumption Chalmers makes, however, is implicit in his characterization of "virtual fictionalism" as the thesis that "virtual worlds are fictional worlds":

> [O]n this view virtual worlds have a status akin to Tolkien's Middle Earth, and virtual objects have a status akin to that of Gandalf or the One Ring: they do not exist in reality, but only in fiction.[11]

The suggestion that virtual objects derive their fictional status from that of the "world" that contains them is starkly idiosyncratic, relative to the explanatory strategies normally used by metaphysical fictionalists about other types of entities. According to the account of pretense that tends to be favored by these philosophers (which they borrow from Kendal Walton's theory of representational art), thought and discourse about fictional objects are occasioned by the discrete mental act of seeing certain isolated aspects of the real world as "props," and hence needn't be prefaced by elaborate acts of imaginative world-building.[12] A speaker might introduce a new fictional object by way of an entirely off-the-cuff "existential metaphor," as Stephen Yablo puts it. Thus, for example, might *gas in the tank* be introduced as a metric for a tired person's remaining energy, or *ducks in a row* as a metric for the items on a daily to-do list, both in a less stable, but otherwise similar fashion, to how *numbers* are invoked as a metric for cardinality. "A metaphor on this view," proposes Yablo (2000),

> Is an utterance that represents its objects as being like so; the way they would need to be to make it pretense-worthy – or more neutrally, sayable – in a game the utterance itself suggests.[13]

The singular utterance, together with whatever ontological implications it appears to have all by itself, *comes first* according to this model; if the fictional objects referred to therein eventually come to seem part of a larger "world," that is an occasional and utterly ancillary feature of the fundamental process.

An instructive analogy can be drawn here with how, in *Naming and Necessity* (1972), Saul Kripke famously describes the relationship between ordinary singular terms with real-world referents such as "Richard Nixon" and the various possible worlds that their referents might also inhabit:

> [A] possible world isn't a distant country that we are coming across, or viewing through a telescope…A possible world is *given by the descriptive conditions we associate with it*…if someone makes the demand that every possible world has to be described in a purely qualitative way, we can't say, 'Suppose Nixon had lost the election', we must say, instead, something like, 'Supposed a man with a dog named Checkers, who looks like a certain David Frye impersonation, is in a certain possible world and loses the election.'…This is just not the way we ordinarily think about counterfactual situations. We just say 'suppose this man had lost'. It is *given* that the possible world contains *this man*, and that in that world, he had lost.[14]

There is considerable *prima facie* appeal to the idea that reference both to fictional objects and to real objects within other possible worlds can be achieved from the *inside out*, as it were, as well as just from the *outside in*. It lowers the bar of success somewhat for our attempt at responding to Stanley's phenomenological objection since it's at least a bit easier to imagine that the exercize of the fictional imagination might occur unconsciously if it doesn't always have to involve the conceptualization of a whole fictional world. It also allows us to think of an agent's cognitive relations to fictional objects as achievable without the exercize of any prior representational capacity (abstraction, visualization, or symbolization, for example). This will bring our account of the fictional imagination into harmony with the types of social, enactive accounts of the nature of pretense that have recently been defended by both philosophers and psychologists.[15] Most importantly, though, it will turn out that some of the cognitive states peculiar to the user of contemporary VR technologies also provide what is perhaps the clearest real-world illustration of this type of bimodality in the operation of the fictional imagination.

Before comparing Chalmers' account of the phenomenology of VR with fictionalism thus conceived, it will be worth making a short digression to examine a recent fictionalist critique of the most basic ontological assumptions Chalmers' type of realism. In a paper from 2019, Neil McDonnell and Nathan Wildman argued that Chalmers' account of the relationship between data objects and the physical world is metaphysically indefensible. They invoke Jaegwon Kim's famous causal exclusion argument against non-reductive physicalism in an attempt to demonstrate that virtual objects cannot have the sorts of causal powers they would need to qualify as real while also depending on their existence upon the properties of data structures. A virtual table, calculator, or monster cannot, they propose, be credited with the capacity to bring about perceptual experiences for a VR user if such entities depend for their own existence upon discrete digital processes and states of the machines that run the relevant applications. For this, they argue, would violate a metaphysical prohibition against causal redundancy that seems to follow from the causal closure of the material world.[16]

Having offered this critique of the realist's view, McDonnell and Wildman proceed to defend the hypothesis that virtual objects are fictional on the grounds that it avoids the problem. The description they provide of how virtual objects arize from acts of pretense allows (at least implicitly) for both of the possibilities I described earlier on. They portray the VR user as being engaged in a game of make-believe wherein she uses images, sounds, or stimuli from haptic feedback devices as props to represent virtual Frisbees, monsters, or whatever. They characterize this activity as being "authorized" by the rules of the relevant game in the same way that our thoughts about characters in a novel are authorized by how they fit into the imaginary universe where the overall narrative takes place. But McDonnell and Wildman also allow that props can, reciprocally, be "used to guide and determine the contents of a particular game of make-believe," since "nearly everything can be an *ad hoc* prop, pressed into service for a single game of make-believe on a single occasion." This latter possibility, they argue, requires us to expand the reference of the term "fiction" to "many things we naturally tend to not call fictions – dolls, dances, and disco songs, as well as novels, paintings, and plays."[17]

In a paper responding to McDonnell and Wildman's criticisms, Chalmers (2022) reiterates his claim that virtual objects are grounded in the mental states of VR users *as well as* in data structures. He describes the virtual object/data structure relation as being analogous to the relationship between a gang and its members, and to the relationship between money and physical tokens. In all such cases, what ensures the continuity of the former, composite entity over time is *both* the continuity of its parts *and* the willingness of thinking subjects to view such objects (a VR Frisbee, a motorcycle club, a dollar) as persisting in spite of changes in underlying physical features.[18] He also points out that, when a VR app is running on two different servers, it should be perfectly acceptable to say that the same virtual entity is grounded in two physically discrete data structures.[19]

Given these considerations, there is no reason to think that the causal powers we might attribute to virtual objects, gangs, or dollars lead to the type of over-determination that Kim's argument targets.

Such observations should be familiar to anyone conversant with the recent literature on the "grounding" relation in analytic metaphysics. Philosophers who believe that it is the grounding relation that obtains in circumstances where others have seen supervenience, reducibility, or mere identity have tended to adopt a "permissivist" approach to questions of existence as applied to whatever types of entities X they describe as being grounded in entities of type Y.[20] Without wading too deeply into this debate, it seems to me that there is nothing sufficiently problematic with Chalmers' account of this type of asymmetric dependence relation between data structures and virtual objects to give fictionalism presumptive credibility all by itself.

Let us return, then, to Chalmers' account of the ways in which virtual objects depend upon our mental activity. As I remarked earlier, he associates fictionalism with the view that our reactions to VR environments (at least when these are at their most immersive) can best be explained *via* the hypothesis that virtual objects generate the illusion that they are real. The VR user's reaction to the flying virtual Frisbee or the pouncing virtual monster would, on this type of account, be analogous to seeing water in a mirage or a pink sky through rose-tinted glasses. Chalmers acknowledges that unpracticed users of VR devices such as the Oculus or Vive might well "viscerally" respond to them in this type of way. But he also claims that the "sophisticated"[21] user recognizes and responds to the relevant objects *as* virtual, and neither behaves nor thinks about the objects of her current perceptions anything like the way she would if she were subject to a mere illusion. For, during the transition from VR naïveté to VR sophistication, what Chalmers refers to as "cognitive orientation" will almost inevitably take place, and the psychologically normal VR user will be bound to acquire the habit of viewing virtual Frisbees and monsters *as* virtual objects.

To support this point, Chalmers develops a suggestive analogy between the experience of VR and the perception of objects in mirrors. Any driver who has become habituated to using a car's rear-view mirror will see the other cars it reflects as being behind her own. Similarly, habituated users of a regular plane mirror will see the objects reflected therein as being on the near side of the glass. This, Chalmers proposes, can be taken as evidence for the plausibility of what he calls the "no-illusion" thesis about "mirror phenomenology."[22] And the case to be made for veridicalism about mirrors is analogous to the case that he thinks can be made for realism about virtual objects, at least once VR users have overcome an initial phase of illusion-prone naiveté. The VR sophisticate will not, of course, think that there are *no* ontologically significant differences between virtual and real-world stones, spaceships, or flowerpots. And Chalmers also takes pains to point out that the question of whether a virtual X also qualifies as an X *simpliciter* will never be entirely trivial, since it will depend upon whether the identity conditions of the relevant type of object are best understood in functional terms (e.g., a calculator) or in terms of their physical constitution (e.g., a rock).[23] But these distinctions are clearly of quite different ontological purport that the distinction between what is real and what is fictional/illusory.

Chalmers' strategy for defending virtual realism by appeal to the phenomenon of cognitive orientation has considerable intuitive credibility. It is certainly the case that most VR users take a while to adapt to the medium, and it seems natural enough to characterize this development in their responses (both cognitive and behavioral) to virtual environments as involving a gradual reduction in their susceptibility to perceptual illusion.

Nonetheless, a curious side-effect of Chalmers' argument is its vulnerability to a criticism analogous to Stanley's phenomenological objection against fictionalism. Recall Stanley's complaint that the fictionalist implausibly characterizes the attitude of pretense as unconscious, in the strong sense of its being *inaccessible* to us (perhaps even in principle). Chalmers' brand of realism has exactly the same sort of problem, at least insofar as the VR "sophisticate" will continue to be *immersed* in her environment. For, to the extent that she finds her virtual environment immersive, she is bound to show proportionately less responsiveness to real-world stimuli and a proportionately greater tendency to navigate virtual stimuli in similar ways to how she would react to their real-world counterparts. And it seems fair to assume that (at least *ceteris paribus*) this will leave her with less capacity to access whatever cognitive states are involved in regarding the virtual objects therein as ontologically distinct from those she interacts with in the non-virtual world, at least for as long as her VR experience continues.

This at the very least shows that there is a significant (albeit subtle) disanalogy between the reasons that support mirror veridicalism and the reasons that Chalmers takes to support virtual realism. For the difference between the true *ingénue* about plane mirrors and someone who has become accustomed to their distinctive characteristics will be bound to issue in gross behavioral differences (e.g., the former might reach forward to grab a reflected object that is actually behind him, bumping his hand on the glass). But the precize character of any behavioral or cognitive shift between the phase during which virtual objects are illusory and the point at which the user supposedly recognizes them as virtual will depend upon ancillary features of both the particular program being run on her device and her degree of involvement with its specific interactive features. In virtual environments that feature a high degree of what Erick Ramirez and Scott LaBarge call *perspectival fidelity*[24] or *context realism*,[25] the user's reactions to perceptual stimuli might both feel to him and seem to an impartial observer very much *as if* the virtual objects he is perceiving are real, even though he might not in any sense qualify as *believing* that they are real.

It is far from clear, then, to what extent Chalmers' account of the mental states of sophisticated VR users is *incompatible* with Wildman & McDonnell's fictionalist proposal that interacting with virtual environments involves an ineliminable element of pretense. And it seems to me that, so long as we focus our attention solely upon the user's occurrent mental states when presented with images of individual virtual objects, we are unlikely to come across any additional decisive piece of evidence that will incline us toward one or the other account of VR phenomenology, let alone settle the metaphysical debate that they are supposed to help one adjudicate. In the next section, I will explore a different strategy for breaking this apparent stalemate, by adopting a somewhat more holistic perspective upon the experience of using VR technologies.

Bimodal Pretense and Lusory Fictionalism

Recall once again the central ambiguity I characterized earlier as being at the heart of the fictionalist project: does the status of individual entities as fictional derive from that of the imaginary "worlds" that they occupy or from some more discrete, object-focused act of the mind? It seems to me that this aspect of fictionalism, while it has generated some confusion elsewhere, is actually more of a feature than a bug when it comes to figuring out the metaphysics of VR.

This is because, at least when interacting with the most common types of VR devices at the present state of the technology (i.e., motion-tracking headsets and associated peripherals such as hand controllers, haptic gloves, omni-directional treadmills, etc.) the user will inevitably undergo two separate, temporally discrete episodes in the course of her immersion into a virtual environment. The types of interactions that have been the focus of almost all the philosophical discussion so far – i.e., when the user sees or otherwise senses a suite of virtual objects and responds to them interactively – can only occur when they are preceded by the phenomenologically distinct moment of transitioning out of the non-virtual world, by interposing these devices between one's sensorium and its more customary sources of input.[26] What is distinctive of the experience of VR during these transitional episodes is not what one perceives; rather, it is what one *decides* and what one *anticipates*. The wearer of a VR headset undertakes a commitment through its use to undermine her normal means of mental access to the objects of her everyday environment, while at the same time knowing (or, at least normally, expecting) that her sensory, reflexive, and cognitive reactions to objects in the virtual world will continue to occur to some significant extent as though the change had never happened.

Different types of VR interfaces can require that this type of commitment be made with widely varying degrees of strength and exclusivity. And one end of the scale would be the types of (thus far merely science-fictional) "neural cut-out" HMDs depicted in Gibson's *The Peripheral,* which would allow for absolutely no residual interaction between VR user's sensorium and his real-world environment. AR lenses that filter out only carefully selected external visual cues or that overlay images onto one's ordinary visual field perhaps represent the opposite extreme. But it is this curious feature of the mind's engagement with VR – the more-or-less radical separation, that is, between an undertaking of preliminary, world-oriented cognitive commitments and the subsequent attitudes taken toward discrete objects that "occupy" a virtual world – that seems to me to provide an especially useful perspective on certain otherwise ambiguous features of the fictionalist project.

What conclusions will we end up reaching if we try to understand the mind's contribution to the metaphysical character of virtual objects by concentrating upon this temporally bifurcated – but in an important sense, unitary – mental reorientation, rather than upon singular acts of *pretense* (as in McDonnell and Wildman's account) or upon the gradual development of *sophistication* (as this is described by Chalmers)? I do not think that this shift in explanatory focus is enough by itself to underwrite an unqualified endorsement of fictionalism about virtual objects. But I shall offer some reasons for thinking that, once we understand VR users' attitudes toward their virtual environments as fundamentally bimodal in this way, it does become possible to identify ontologically significant differences between every day and virtual objects that arize from other factors than just differences in their physical grounding outside of the mind.

Recall that, for Chalmers, the crucial point worth making against "virtual fictionalism" is that the perception of VR environments is not best understood as illusory. An illusion is a fundamentally *deceptive* experience. Except in relatively rare (or science-fictional) circumstances where a VR user is somehow tricked into donning the headgear, gloves, etc., it would be indeed be perverse to insist that her performance of this action was an act of outright self-deception, analogous to submitting to hypnosis or succumbing to delusionally wishful thinking. That having been said, however, there certainly is something about voluntary VR use that involves the deliberate *undermining* of one's own ordinary cognitive processes. As

Michael Heim observes in one of the earliest philosophical works written explicitly about VR, a partly constitutive feature of the technology it relies upon is that it has the power to update information presented to the user's sensorium "as fast as the human organism can alter her physical position and perspective."[27] The state in which one's normal mechanisms for perceptual response are subjected to this type of manipulation must therefore lie somewhere upon an epistemic continuum that extends from radical, Cartesian misapprehension of the world at one end to a gentler type of complicity with false representations at the other – the sort that one entertains, perhaps, when fantasizing about being in Middle Earth, an unvisited foreign country, or some remembered environs from one's own past. One might even shift back and forth between the two extremes during a single session of VR, though this would probably be more likely to happen during the use of AR devices that integrated simulated experiences relatively seamlessly into one ordinary interaction with the physical world. Gibson's phrase "consensual hallucination" has often been invoked in this sort of context,[28] and it does seem to capture something important about the distinctive conditions one chooses to enter when first interacting with VR technologies.

Although not all VR apps are conventionally classifiable as games,[29] a provocative analogy can be drawn between the type of cognitive reorientation imposed by contemporary VR interfaces and the kind of temporary (but often quite radical) axiological reorientation that happens during gameplay. This type of shift in evaluative attitudes that agents voluntarily undergo when they choose to play games receives its most philosophically subtle exposition in Bernard Suits' classic 1978 study *The Grasshopper*. In response to the famous Wittgensteinian thesis that "game" is undefinable,[30] Suits describes a particular mental orientation (the "lusory attitude") that he argues is fundamentally constitutive of gameplay. An agent who adopts the lusory attitude decides to pursue some goal or other in a fundamentally inefficient manner, in the specific sense that she follows rules (i.e., those of the relevant game) that make the goal more difficult to achieve than it would be otherwise. [31] The bowler chooses, for example, to knock down wooden pins with a ball from a far distance rather than simply walking down the lane and kicking them over; the ringtaw player chooses to acquire a rival's marbles by knocking them out of a circle, rather than simply grabbing them and running. Some of the actions performed while playing either of these games thereby take on a special type of double-sided teleology: the player is, on the one hand, certainly still *pursuing* the goals just mentioned, but not because he values their *achievement* so much as because he values the *experience of their pursuit* under the relevant set of restrictions. The axiological commitments that the gamer undertakes are in an important sense isomorphic to the cognitive states adopted by the VR user toward specific objects of perception once she has put on the goggles, gloves, or whatever. In both cases, the fundamental character of occurrent mental processes (aiming at the glass marble; glimpsing the approach of the virtual Frisbee) is partly constituted, not by any contemporaneous feature of the mind's activity, but by discrete commitments undertaken at an earlier time (to play the marbles game; to put on the headset).

Does it seem fair to say that the game-player's in-game goal of knocking down the pins or of acquiring the marbles is not his *real* goal, on the grounds that his pursuit of it is an accidental feature of the experiential preferences he has manifested earlier by entering the game in the first place? To the extent that it does, the analogy I have just drawn with VR suggests we would be proportionately justified in saying that the virtual Frisbee is also unreal, in spite of the fact that one's spontaneous, "sophisticated" response to its approach might closely resemble how we normally react to an approaching projectile out in the non-virtual world.

These observations bring us back to Stanley's phenomenological objection to fictionalism. Stanley's central concern, once again, was that the fictionalist about some type of object X must attribute a radical failure in first-person authority about the content of our attitudes toward Xs. But if virtual objects can plausibly be characterized as fictional for the reasons just given, this turns out to be because of the existence of an *intertemporal relation* between the VR user's thoughts while confronting the virtual object and her original entry into VR, rather than on account of the content of whatever unconscious, subconscious, or sub-doxastic states happen to accompany her ducking of the virtual Frisbee or her punching virtual buttons on the virtual calculator. And the extent to which this relation is accessible to her at that specific point in time might depend upon all sorts of utterly extraneous factors - e.g., the limits of her short-term memory, the depth of her immersion in a particular environment, or the perceptible resemblance of that environment to whatever her environment was before she plugged in. Of course, the relation will only be *inaccessible in principle* to the VR user in certain types of liminal, sci-fi scenarios, such as the choice to plug into a Nozick-style "experience machine" or the central decision made by the protagonist of the 2001 film *Vanilla Sky*. But these considerations are surely still enough to undermine whatever credibility Stanley's claims about first-person authority might otherwise have possessed as an objection to fictionalism about virtual objects.

Does this partial vindication of fictionalism about virtual objects have anything to teach us that might help support fictionalist metaphysics more generally? It does seem to me that the strategy I have used to disarm the phenomenological objection to hermeneutic fictionalism might generalize in an interesting way. For, perhaps, when we speak about numbers, knowledge, universals, or instances of moral goodness, we lack first-person access to the element of pretense in our speech because it derives from some earlier (partly or wholly unremembered) moment of fundamental reorientation toward the world analogous to the type of reorientation that takes place when one first slips on a VR headset. In some cases such an originary undertaking might have been made by us, ourselves who currently use the relevant vocabulary; in others, it might be more credibly attributable to the linguistic innovators whose usages we have learned to mimic.[32] And given the continuing ubiquity of "language game" talk in contemporary philosophy, there is surely substantial precedent for extending the analogy between VR and gameplay to even our most superficially earnest, literally-intended everyday speech about the world around us.[33]

Notes

1 See, e.g. Paul Spade, trans., *Five Texts on the Mediaeval Problem of Universals: Porphyry, Boethius, Abelard, Duns Scotus, Ockham* (Indianapolis: Hackett, 1994), and Berkeley, G. (1710/1982). A treatise concerning the principles of human knowledge. Indianapolis IN: Hackett Publishing.

2 See, e.g. C.L. Stevenson, "The Emotive Meaning of Ethical Terms," *Mind,* 46: 181 (1937), pp. 14–31 and Simon Blackburn, *Essays in Quasi-Realism* (New York: Oxford University Press, 1993).

3 The former, broader characterization of fictionalism can be found in Mark Kalderon (ed.), *Fictionalism in Metaphysics* (Oxford: Clarendon Press, 2005), p. 1 and Mark Sainsbury, *Fiction and Fictionalism* (London: Routledge, 2009), p. 139.

4 See Jason Stanley, "Hermeneutic Fictionalism," in H. Wettstein (ed.), *Midwest Studies 25: Figurative Language* (Malden, MA: Blackwell, 2001), p. 36.

5 Stanley, p. 46.

6 See, e.g., John p. Burgess, "Fictionalism as a Phase (to be Grown Out of)," in Bradley Armour-Garb and Frederick Kroon (eds.), *Fictionalism in Philosophy* (New York: Oxford University Press, 2020), pp. 48–60 and Stephen Yablo, "Apriority and Existence," in p. Boghossian and C. Peacocke (eds.), *New Essays on the a Priori* (Oxford: Oxford University Press, 2000), pp. 197–226.

7 See Julianne Chung, "Could Knowledge-Talk be Largely Non-Literal, *Episteme,* 15: 4 (2018), pp. 383–411 and Mark Eli Kalderon, "Moral Inference, the Frege-Geach Problem, and Reasonable Inference," *Analysis,* 68: 2 (2008), pp. 133–143.

8 See Grant Tavinor, *The Art of Videogames* (Chichester: Wiley-Blackwell, 2009), pp. 51–52, and James J. Cummings and Jeremy N. Bailenson, "How Immersive Is Enough? A Meta-Analysis of the Effect of Immersive Technology on User Presence," *Media Psychology,* 19: 2, (2006) pp. 272–309.

9 In *Reality+* (New York: W.W. Norton & Company, 2022), at pp. 124–128, Chalmers discusses (and is receptive to) the further suggestion that all our perceptions of the physical world might also be grounded in data structures, if Nick Bostrom's (2003) simulation hypothesis is true. See Nick Bostrom,

"Are You Living in Computer Simulation?" *Philosophical Quarterly,* 53: 211 (2003), pp. 243–255.

10 See Chalmers, "The Virtual and the Real," *Disputatio,* 9: 46 (2017), p. 349, fn. 16 and 17.

11 Chalmers, "The Virtual and the Real," p. 315.

12 For more on this topic, see Grant Tavinor, "Against Metaphysical Interpretations of VR," this volume.

13 Yablo, "Apriority and Existence," pp. 213–214.

14 Saul Kripke, *Naming and Necessity* (Cambridge, MA: Harvard University Press, 1972), p. 44–46.

15 See, e.g., Zuzanna Rucińska, "Social and Enactive Perspectives on Pretending," *Avant,* 10: 3 (2019), pp. 1–27, and Somogy Varga, "Pretense, Social Cognition and Self-Knowledge in Autism," *Psychopathology,* 44: 1 (2011), pp. 46–52.

16 See McDonnell and Wildman, "Virtual Reality: Digital or Fictional?" *Disputatio,* 9: 55 (2019), pp. 383–386.

17 McDonnell and Wildman, pp. 390–391.

18 See Chalmers, "The Virtual as the Digital," *Disputatio* 11: 55 (2019), p. 458.

19 Chalmers, "The Virtual as the Digital," p. 465.

20 Two of the most widely discussed such relations, as identified by Jonathan Schaffer in his seminal paper, "On What Grounds What?" in David Chalmers, David Manley and Ryan Wasserman (eds.), *Metametaphysics* (Oxford: Clarendon Press, 2009), pp. 347–383] are the relation between composite objects and their proper parts and the relation between mind and brain (see pp. 358–363). It is worth noting that Schaffer also adopts a permissive attitude toward the existence of fictional objects themselves: metaphysical disputes about God or Santa Claus should, he suggests, be concerned with their *fundamentality,* rather than their existence. Although I have myself associated fictionalism with irrealism and anti-realism here, none of the arguments made in the present paper are strictly incompatible with such an approach.

21 Chalmers, "The Virtual and the Real," p. 327.

22 Chalmers, *Reality+,* pp. 213–214.

23 Chalmers, "The Virtual and the Real," pp. 324–325.

24 I.e. "the Degree to Which a Representation Accurately Depicts the Subjective Point of View of a Neurotypical Human Being." See Erick Jose Ramirez and Scott LaBarge, "Real Moral Problems in the Use of Virtual Reality," *Ethics and Information Technology,* 20 (2018), p. 255.

25 I.e. "the more a virtual world's environment is bound by the same physical and psychological principles that a subject believes grounds their own world, and the more these rules cohere with a user's lived experience." See Ramirez and LaBarge, "Real Moral Problems in the Use of Virtual Reality," p. 255.

26 A consequence of this proviso is that the conclusions reached in the present paper should not be thought to apply (at least without further qualifications) to possible future VR devices that might not require users to pass through a such temporally discrete transitional stage, or to environments that can transition back and forth between virtual and augmented reality, or to the type of VR that we would all be experiencing if Nick Bostrom's Simulation Hypothesis happened to be true [see Nick Bostrom, "Are You Living in a Computer Simulation?" *Philosophical Quarterly,* 53: 211 (2003), pp. 243–255].

27 Michael Heim, *Virtual Realism* (New York: Oxford University Press, 1998), p. 7.

28 E.g. in Chalmers, *Reality+*, p. 297.
29 This is a point that Chalmers feels it necessary to make much of in "The Virtual as the Digital" (see p. 484).
30 See Ludwig Wittgenstein, *Philosophical Investigations*, Third Edition, Trans. Elizabeth Anscombe (New York: Macmillan, 1968), p. 31.
31 See Bernard Suits, *The Grasshopper: Games, Life, and Utopia* (Boston: David R. Godine, 1990), pp. 35–41
32 Richard Joyce has suggested, in a similar vein, that references to some types of fictional entities have a similar epistemic profile to so-called "dead metaphors." To the extent that this analogy is plausible, achieving introspective access to one's own fictionalizing attitudes will face the same type of difficulty as determining whether expressions such as "the mouth of the river" or "face of a clock" still qualify as metaphorical at all. See Richard Joyce, "Fictionalism: Morality and Metaphor," in Bradley Armour-Garb and Frederick Kroon (eds.), *Fictionalism in Philosophy* (New York: Oxford University Press, 2020), pp. 113–116.
33 Earlier versions of this paper were delivered to audiences at the 2018 Workshop oi Virtual Reality at ITU Copenhagen and the 2019 annual meeting of the American Society for Aesthetics in San Antonio. I am grateful to Paal Antonsen, David Chalmers, John R. Sageng, and Karolina Wisniewska for helpful comments and discussion.

References

Berkeley, G. (1710/1982). A treatise concerning the principles of human knowledge. Indianapolis IN: Hackett Publishing.

Blackburn, S. (1993). *Essays in Quasi-Realism*. New York: Oxford University Press.

Bostrom, N. (2003). Are you living in a computer simulation? *Philosophical Quarterly* 53(211): 243–255.

Burgess, J.P. (2020). Fictionalism as a phase (to be grown out of). In *Fictionalism in Philosophy*, Bradley Armour-Garb and Frederick Kroon, eds. New York: Oxford University Press: 48–60.

Chalmers, D. (2022). *Reality+*. New York: W.W. Norton & Company.

Chung, J. (2018). Could knowledge-Talk be largely non-literal? *Episteme* 15(4): 383–411.

Heim, M. (1998). *Virtual Realism*. New York: Oxford University Press.

Joyce, R. (2020). Fictionalism: Morality and metaphor. In *Fictionalism in Philosophy*, Bradley Armour-Garb and Frederick Kroon, eds. New York: Oxford University Press: 113–116.

Kalderon, M. (2005). *Fictionalism in Metaphysics*. Oxford: Clarendon Press: 1.

Kalderon, M.E. (2008). Moral inference, the Frege-Geach problem, and reasonable inference. *Analysis* 68(2): 133–143.

Kripke, S. (1972). *Naming and Necessity*. Cambridge, MA: Harvard University Press.

McDonnell, M. & Wildman, M. (2019). Virtual reality: Digital or fictional? *Disputatio* 9: 55.

Ramirez, E.J. & LaBarge, S. (2018). Real moral problems in the use of virtual reality. *Ethics and Information Technology* 20: 249–263.

Rucińska, Z. (2019). Social and enactive perspectives on pretending. *Avant* 10(3): 1–27.

Sainsbury, M. (2009). *Fiction and Fictionalism*. London: Routledge: 139.

Schaffer, J. (2009). On what grounds what? In *Metametaphysics*, David Chalmers, David Manley and Ryan Wasserman, eds. Oxford: Clarendon Press: 347–383.

Spade, p. (1994). *Five Texts on the Mediaeval Problem of Universals: Porphyry, Boethius, Abelard, Duns Scotus, Ockham. (Translator)*. Indianapolis: Hackett.

Stanley, J. (2001). Hermeneutic fictionalism. In *Midwest Studies 25: Figurative Language*, H. Wettstein, ed. Malden, MA: Blackwell: 36.

Stevenson, C.L. (1937). The emotive meaning of ethical terms. *Mind* 46(181): 14–31.

Suits, B. (1990). *The Grasshopper: Games, Life, and Utopia*. Boston: David R. Godine: 35–41.

Tavinor, G. (2009). *The Art of Videogames*. Chichester: Wiley-Blackwell.

Varga, S. (2010). Pretense, social cognition and self-knowledge in Autism. *Psychopathology* 44(1): 46–52.

Wittgenstein, L. (1968). *Philosophical Investigations*, Third Edition, Trans. Elizabeth Anscombe. New York: Macmillan: 31.

Yablo, S. (2000). Apriority and existence. In *New Essays on the a Priori*, p. Boghossian and C. Peacocke, eds. Oxford: Oxford University Press: 197–226.

3 Against Metaphysical Interpretations of VR

Grant Tavinor

The Conceptual Framing of VR

The increased philosophical prominence of virtual reality, which has resulted in a torrent of new papers and books on the topic, is principally the result of technological and media developments in the last decade. Though artificial computer realities have been conceived since at least the 1970s (Krueger, 1983) it is only recently that virtual reality experiences have had a widespread availability for the public. Devices such as the Oculus Rift and PlayStation VR have now found their way into millions of homes and the ubiquitous image of a person wearing a VR headset and directing their attention to "somewhere" other than their immediate environment, is a cultural staple. The popularity of the term "virtual reality," is often attributed to Jaron Lanier, who not only coined the term in the 1980s during his work on games systems, but also popularised it through his early activism (Conn et al., 1989). Alternative terminologies for VR have appeared and had a certain amount of influence, including "artificial reality" (Krueger, 1983), "cyberspace" (Gibson, 1984), "synthetic environments" (Gigante, 1993), "synthetic worlds" (Castronova, 2005), and "mixed reality" (Milgram & Kishino, 1994). Historical accident might have seen any of these taking prominence, but for whatever reason, whether Lanier's efforts, or the memorable initialism "VR," "virtual reality" is now the dominant term.

Note however, that what "virtual reality" shares with these alternative terminologies is that they are all framed around the representational content typically realised by the new technologies. Each refers to the apparent alternative places that VR allows us to encounter or visit, and it may be the sense of ontological mystery aroused by this "transportive" or "immersive" potential that has secured the dominance of this conceptualisation. Alternative realities where one can experience different worlds or live different lives have had a similar prominence in the popular understanding of VR, particularly in the form of the science fiction premises explored in works such as *Star Trek: The Next Generation* (1987–1994) and *The Matrix* (1999).

DOI: 10.4324/9781003359494-5

And yet, fundamentally, VR comprises a collection of largely novel media technologies including stereoscopic head mounted displays, haptic and force feedback gloves, binaural audio, and kinaesthetic devices that allow for the representation of a user's movement through space. Not only do these technologies have non-VR precedents—in the case of stereoscopic headsets, the children's toy *View-Master* easily comes to mind—but their representational nature can be framed and understood in terms of theories we might apply to other forms of representational practices and artifacts. Particularly, the *depictive* principles underlying stereoscopic headsets can be linked historically to the developments in depiction involved in the graphical realisation of naturalistic spaces. To do so, we might reach out to the fields of art history and philosophical aesthetics for the concepts and traditions of theory developed there. Taking this conceptual approach might lead to a very different kind of terminology for what we now refer to as virtual reality.

This accident of nomenclature has had a significant effect on how the phenomena under discussion elsewhere in this collection has been conceived and theorised by scientists and philosophers. The framing of these new media developments around the apparent representational content typical of their use—alternative worlds, environments, and objects—and particularly, the reference to these phenomena as a kind of modified reality, has naturally tempted theorists to adopt a metaphysical framework in their accounts of VR, an approach that frequently abstracts away from questions of how such representational content is technologically realised. Virtual reality is thus to be understood in *ontological* terms, and the natural question to ask of VR is of the metaphysical nature of the worlds, environments, and objects it presents. An early source of the metaphysical orientation is Michael Heim's various writings on the metaphysics of virtual reality (Heim, 1993; Heim, 1998). The ontological framing of the new medium also came at a time when philosophical scepticism about a simple reality was undergoing a seeming renaissance (Goodman, 1979; Baudrillard, 1994), and this is perhaps part of the attraction of alternative realities in some academic circles.

This framework has been prominent in scientific accounts of VR, and perhaps, supplies the dominant conceptual schema currently applied to VR and related technologies in the form of the "mixed reality spectrum" or "virtuality continuum" (see Figure 3.1)

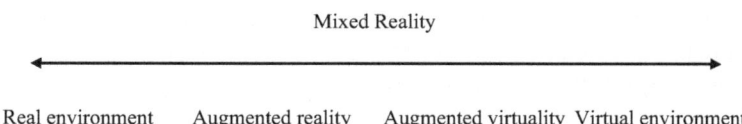

Mixed Reality

Real environment   Augmented reality   Augmented virtuality Virtual environment

Figure 3.1 "The mixed reality spectrum," Adapted from Milgram and Kashino, (1994).

In their influential work, VR theorists Paul Milgram and Fumio Kishino characterise the difference between electronic displays such as conventional video displays, and VR displays, as dually defined by these media differences, and the nature of the objects and environments they inherently represent (1994). In describing a now familiar visualisation of the continuum, they argue that,

> The concept of a "virtuality continuum" relates to the mixture of classes presented in any particular display situation [...] real environments, are shown at one end of the continuum, and virtual environments at the opposite extremum. The former case, at the left, defines environments consisting solely of real objects [...] and includes for example what is observed via a conventional video display of a real-world scene. An additional example includes direct viewing of the same real scene, but not via any particular electronic display system. The latter case, at the right, defines environments consisting solely of virtual objects [...], and example of which would be a conventional computer graphic simulation.
> (Milgram & Kishino, 1994)

Here, displays are tied very closely to the ontological classes of objects they represent, in this case real or virtual, and a mix between these. Following this lead, characterizing VR as involving a distinctive class of virtual objects and environments is now relatively widespread in the social science, psychological, and technological literature on VR. Skarbez, Smith, and Whitton (2021) review the extensive impact the idea of a reality-virtual continuum has had on the VR literature, but also note some of the limitations that over time have become evident in the conceptual framework.

Philosophers have also naturally found this framing amenable to the traditional metaphysical and epistemological concerns of their discipline. Scepticism about the reality of our own current phenomenal world, and what this might say about the idea of reality itself, have long been standard concerns of philosophy: the early scepticism of Plato's cave prefigures both Descartes' methodological scepticism and Berkeley's idealism, and all find a recent form in Nick Bostrom's simulation argument (Bostrom, 2003). These ideas arise almost inevitably from the new technological developments of virtual, unreal, or simulated worlds. So, for example, in his account of the metaphysics of virtual reality, Phillip Brey identifies two ontological questions as his principal concern

(1) What is the mode of existence of virtual objects, actions, and events?
(2) Can any virtual objects, actions, or events be claimed to be part of

the real world as opposed to being unreal, a merely simulated reality, and if so, how does this fact problematize the distinction between reality and virtuality? (Brey, 2014: 43)

In his well-reasoned answer to these questions, which makes use of familiar ideas from the ontology of social kinds such as money, Brey argues that unlike physical reality, which can only be simulated in VR, "institutional reality can in large part be *ontologically reproduced* in virtual environments" (Brey, 2014: 47). This, of course, has been the experience of many of us, as even pre-pandemic so many institutions and practices that previously had a brick-and-mortar existence, were *virtualised* when they migrated to digital platforms.

This ontological approach to VR often centres on the question of whether virtual worlds and objects are *real* or *fictional* things, perhaps the most prominent conceptual distinction on VR within its recent philosophy.[1] David J. Chalmers is notable for contrasting his realist account of virtual objects with what he calls "VR fictionalism," arguing that virtual objects have causal and perceptual propensities that fictional objects cannot have (Chalmers, 2017: 317–320). He subsequently names his position on the status of the objects encountered in VR—both causally and perceptually—as "digitalism," stating quite forcefully that "virtual reality is a sort of genuine reality, virtual objects are real objects, and what goes on in virtual reality is truly real" (Chalmers, 2017: 309). Chalmers' account exemplifies the metaphysical approach to VR by turning immediately to the ontological question of whether VR objects and environments are real or merely fictional things, whether experiences in VR are illusory, and whether VR experiences are as valuable as real experiences.

Such ontological views may also have a related doxastic component, perhaps further distancing VR from the category of fiction. For some theorists, not only do virtual objects exist, but they generate existential beliefs in those who encounter them. For example, Pavel Grabarczyk and Pokropski characterise the perception of VR worlds and objects in terms of an "ontological shift" (2016) when they argue,

[...] That the obvious prerequisite to being immersed or present in a virtual environment is to *believe* that there is some kind of alternative place in which we can immerse ourselves or be present (as opposed to a simple picture or an animation we can only look at from the outside) [...] not only does the user have to believe in an explorable space, but also that the avatar she explores the space through remains embedded in this space as one of the objects in it.

(Grabarczyk & Pokropski, 2016: 35–36, *my emphasis*).

Here, not only are the objects and environments the appropriate objects of beliefs, but so is the perceiving subject. Realism about virtual perceivers is another frequent component of ontological accounts of VR (for example, see Gualeni & Vella, 2020).

It should be noted that a characteristic choice of examples of virtual reality aids this metaphysical framing. Rather than the limited and often clunky VR technology that currently exists, philosophers often focus on hypothetical and science fiction cases of so-called "perfect" virtual realities (Chalmers, 2003). A perfect virtual reality is one that "from the inside" is indistinguishable from reality, an early example being Robert Nozick's hypothesised "experience machine" (Nozick, 1974) and perhaps the most frequent example being the previously mentioned *The Matrix*. Idealised perfect realities, unlike current VR technology with its limited sensory channels, and in which an awareness of the artificial nature of the VR environment is unavoidable, is ideal for provoking or explaining traditional philosophical scepticism about reality.[2] This repeated focus tempts that conclusion that the *technology* of VR is not of intrinsic interest to most philosophers because it does not generate the metaphysical puzzles and concerns that are their bread and butter.

The Problems with Metaphysics

The dominance of this ontological framing, the prominence it gives distinction between the fictional and the real, and the focus on perfect virtual realities, has been an unfortunate development for our understanding of VR media. Most practically, perhaps, metaphysics has the effect of capturing the debate in the philosophy of VR. I've presented and attended numerous seminars and conference sessions on VR, and a repeated and frustrating experience has been to observe a promising discussion, framed around the aesthetics of VR media, being hijacked by the ontological issues expressed above. A discussion on whether "da Vinci stereopsis" lends objects in VR worlds such as apples a realistic impression impossible in the traditional depictive arts is likely to prompt the question of *what the apple is anyway*. This personal observation is not a philosophical argument against the soundness of the ontological approach, of course, but is in a very real sense a way in which a preoccupation with metaphysics can occlude other promising lines of enquiry, a trend that is not exclusive to the aesthetics of VR. The distracting effects of what at a given time is considered a "serious" concern can be observed throughout the history of philosophy, a classic example being how the austere logical positivism of the early 20th Century eclipsed and set back the of developments of ethics, moral philosophy, and indeed, aesthetics.

In addition to this general worry that metaphysics obscures other approaches, the specific accounts detailed have problems which might partly explain the distracting effects of VR ontology. I've covered some of these difficulties elsewhere, but to give an impression here of how metaphysical approaches can confuse matters, it is worth digging further into an example already mentioned above. Phillip Brey's account of the ontology of VR does an admirable job of disentangling some of the complications of the recent virtualisation of money and social practices such as virtual meetings, but it is still problematic because the account tends to conflate these issues with the narrower concern with the apparent ontology of the objects seen through stereoscopic headsets (Tavinor, 2021: 145–148). That these are two different issues can be seen by examining some of Brey's examples in more detail, particularly, virtual money, virtual apples, and virtual circles. As noted earlier, Brey's ontological conclusions follow from his observation that "institutional reality can in large part be ontologically reproduced in virtual environments." Money, for example, is only "contingently physical" because the features that allow for its function—fungibility, discreteness, scarcity—can be embodied in both physical and digital artefacts (Brey, 2014: 46). An apple, on the other hand, seems *essentially physical*, because apples are fruits with mass, extension, and nutritious properties, and these properties cannot be manifested in a digital or virtual way. Thus, for Brey, any apples that we encounter in virtual worlds are merely "simulated apples," a claim that tempts a metaphysical reading (Brey, 2014: 43).

The shift between virtual money and virtual apples seems to be a subtle change of topic, however. Virtual money typically isn't represented graphically through virtual media such as stereoscopic headsets, and for virtual money to really exist it is sufficient that the fungibility, discreteness, scarcity key to its functioning are maintained through means of computer code, and this is something that can be achieved without a graphical rendering of any kind. Apparent virtual apples, however, are inherently *graphical artefacts*. Leaving aside the further problem that Brey's example of a virtual world is the non-VR world simulator *Second Life*, when a player encounters an apple in the VR version of *Minecraft*, what this means is not that the player has encountered an apple, simulated or not, but that the VR graphics have generated an image with an apple-like appearance.[3] This object is no kind of apple, simulated, virtual or otherwise, rather, it is a *depiction* (even if as in the case of *Minecraft*, it is a depiction that is used to generate the *fictional truth* that one has encountered an apple).

The idea, crucial to metaphysical accounts of VR, that VR necessarily simulates or remediates the apparent objects and environments encountered in VR, is, I have argued elsewhere, a fundamental fault of such approaches (Tavinor, 2021: 146–151). The example of the virtual circle will further illustrate why. Brey argues that

> When in a virtual environment a circle is drawn, the result is a real circle, since a circle is mathematically defined as a phenomenon consisting of points on a plane, and is not by definition a physical object with weight and mass.
>
> (Tavinor, 2014: 46)

While it is true that the representation of a circle in a virtual environment might occasionally involve the production of a circle on a VR display, this will only occur when the apparent position of the virtual camera which defines the perspective in the depiction is precisely perpendicular to the face of the depicted circle. Typically, when a circle or circular object is depicted as a feature of a virtual world, what is likely to be presented on the display is a *projected ellipse*, which is perceived by the viewer as a circle viewed from a distinctive perspective. As such, in its production of a circle in a virtual environment VR media *might never produce an actual circle* (and one should be careful to realise that the set of coded mathematical coordinates that are used by the graphics algorithm to produce the depiction of a circle, also do not comprise a circle).

A further terminological problem with "virtual reality" has now come into view. The term itself is vague, being used in the literature to refer to *at least* three different phenomena, with the frequent assumption that the varied usage is in fact justified by the metaphysical idea that each use refers to the remediation of some customary item in a digital medium. First is the virtualisation of social practices and items such as teaching, money, and shopping. As Brey notes, digital media can ground these practices because the social practices are physically contingent, and because they depend on the conventional understanding and expectations of a social group. Online shopping via digital platforms such as Amazon has thus both revolutionised retail because it sustains many of its fundamental social functions—selection, purchasing, and distribution of goods—but it has also decimated the physically contingent brick-and-mortar aspects of retail, gutting many previously rich social landscapes of consumption and contributing to the phenomenon of the "dead mall." Second, though, are the digital graphical worlds found in *Second Life* and any number of videogames. These *graphical and interactive worlds*, though having their roots in non-digital media—miniature war games, *Choose Your Own Adventure* and *Fighting Fantasy* books, and table-top role-playing games such as *Dungeon and Dragons* share many of the interactive features of virtual worlds—have exploded in number and sophistication since the rise of home computer technology in the 1970s and 1980s. The enormous, and still underappreciated, phenomenon and culture of videogames is the best example of how such digital worlds have now come to dominate the entertainment landscape. Such virtual worlds may depict fantastical or commonplace environments, but they partially overlap with the first sense of virtuality because these graphical environments—and more importantly, the suite of interactive computer programs they combine—are able to sustain social practices previously found in the actual world. Thus, one may chat with friends or play games with them in *Rec Room* because appended to this graphical world is a chat function and various videogames. Finally, some of these graphical worlds are presented through virtual reality media such as stereoscopic headsets or binaural headphones, such as *Rec Room* when it is played on the Oculus Rift. This more limited phenomenon may give a sense of presence to the "inhabitants" of virtual worlds, and because it frequently excludes awareness of a user's actual environment, gives a strong sense that one is experiencing an alternative reality. Philosophers are not always careful to distinguish these three phenomena (Chalmers, 2017: 316; Ludlow, 2017: 17), and when the additional assumption is made that they all involve the ontological remediation of the apparent objects being encountered, the result can only be confusion.

Practically, metaphysical accounts of VR do not connect well with the concerns of the scientists and technologists who are engaged in the research and design of the technology itself. As such, it may provide a very poor foundation for the understanding of the various uses and effects that are being made of and ascribed to VR. For example, VR has increasingly been claimed to have therapeutic uses, including the treatment of anxiety through virtual exposure therapy (Anderson et al., 2013), the treatment of phobias, including "fear of flying, spiders, and heights, as well as panic disorder, agoraphobia, social anxiety, and post-traumatic-stress-disorder" (Jiang, Upton and Newby, 2020: 637), the rehabilitation of brain injuries (Aulisio, Han & Glueck, 2020), Parkinson's disease (Dockx et al., 2016) and strokes (Laver et al., 2017). Metaphysical speculations do little to inform these uses, and indeed may introduce a lack of clarity that hinders our ability to understand and assess them. Recently, medical researchers have bemoaned the lack of definition of VR, a lack which seems partly due to problematic metaphysical assumptions in many proposed definitions (Kardong-Edgren et al., 2019).

Finally, I worry that virtual realism refers, in its more extravagant forms, to *actual idealism*. Nick Bostrom's extraordinary arguments give one of the best examples of this trend. In his now famous paper, Bostrom gives no good empirical reason for the material claim that out apparent world is an "ancestor simulation" (Bostrom, 2003). Rather, the argument comprises an almost narcissistic fantasy about the statistical likelihood that anthropomorphised aliens would choose to simulate the life that we currently happen to be living. Not only is there something both childish and world-denying about such techno-idealism, but this inward-looking subjectivity also seems axiomatic to VR metaphysics. Much of the current philosophical analysis of VR is framed from the perspective of the user's experience and what it feels like from the *inside of the experience machine*. Allied with this, is that the image of VR chosen is so often that of the perfect virtual realities of science fiction, rather than the imperfect virtual realities of current technology. From this perspective, it is the nature of the users' experiences that establishes the reality or lack of reality of VR and its objects and environments. If a user can't tell the difference between virtual and actual reality, then surely, the argument goes, the virtual experience might as well be treated as reality. This argument is the basis for some very strong metaphysical claims (e.g. Chalmers, 2003). But this subjective standard of the reality of VR ignores the function and intentions behind the construction and use of VR technology and its representational content. Inspected "from the outside," the individual's experience within the experience machine counts for little, and in the case that they can't tell the difference between reality and the virtual world that has been constructed for them, the reasonable conclusion is that a person is merely confused or deceived by their world's apparent features, a phenomenon that is demonstrated again and again by the untrustworthiness of perception and those who seek to deceive us.

VR Picturalism

Beyond metaphysical speculations, there is plenty of philosophical interest in VR media. I have argued that the terminology of "virtual reality" has naturally led to questions about the ontology of virtual environments, objects, and perceivers, and to the exclusion of other concerns. However, an alternative terminology and framework might be used both to clarify some of the current confusion surrounding VR, and also reveal previously obscured issues. The conceptual framework I have in mind here—which we might, with the reservation that it is limited to the visual aspects of VR, refer to as *VR picturalism*—is of the art historical practice of relating the viewers of pictures to the space and subject matter depicted within them. The realisation of the impression of space on the 2D surface of a sketch or painting has played a significant role in art history, involving techniques such as linear perspective, *trompe l'oeil*, and the implied presence of the picture viewer in the space of the depicted scene, among others. For example, the philosopher of the arts Richard Wollheim notes that in paintings such as Velasquez' *Las Meninas* and van Eyck's *The Arnolfini Marriage* "the suitable spectator is offered a distinctive form of access through the presence in the represented space—though not in that part of it which is represented—of a figure, whom I call the 'Spectator of the Picture'" (Wollheim, 1998: 225). Through such techniques, viewers of artworks can be given the sense of standing before a space that projects away from them, and which they occupy: indeed, in *trompe l'oeil* pictures the impression may be so strong that the viewers can become confused by what it is they are seeing, perhaps momentarily confusing the picture for reality (Feagin, 1998: 237).

Nevertheless, these attempts at placing the viewer in the picture space are problematic and limited in various respects. Nelson Goodman points out the linear perspective, developed by Brunelleschi and Alberti in the 15th Century as a contribution to the naturalistic depiction of spatial experience, is artificially constrained. For a picture in linear perspective to resemble the reality of its subject, Goodman argues, it must be viewed "through a peephole, face on, from a certain distance, with one eye closed and the other motionless"—and that these constraints on viewing specify "conditions of observation [that] are grossly abnormal" (Goodman, 1976: 13). Moreover, the apparent space in such artworks is insensitive to the changing perspective of picture viewers. In *Las Meninas*, the lady in waiting Isabel de Velasco obscures figures and detail in the background. If the painting really did incorporate the viewer in its space, it would respect the principles of visual parallax, so that by moving before the picture, the viewer could reveal what was previously occluded by the figure.

Of course, *Las Meninas* is a fixed image and so the effects of parallax are impossible to render, but a truly "egocentric picture" would be one where the viewer's changing position in front of the picture did have such perspectival effects. And this is precisely what we find in the visual technology of VR stereoscopic headsets. This is not the only contribution that VR makes to the naturalistic depiction of space, but it is this depictive potential which means that that VR can be best understood as a technological response to the limitations of previous modes of spatial depiction, and the effects for which produces, particularly the sense of "presence" in an alternative space (Tavinor, 2021: 121). The conceptual account that I present here, then, is of VR as a kind of picturing device, specifically, as a form of *egocentric picturing*. It is that this conceptual framework solves or obviates many of the puzzles and confusions involved in ontological accounts of VR that stands as one of its principal virtues.[4]

A little more obviously needs to be said about how pictures can represent or depict apples, circles, and other items without *reproducing* them in an ontological manner. Though I can only give a rough snapshot of the account here, the best way to achieve this is to examine one theory of pictorial seeing, and how VR might be accommodated by the theory. A classic statement of the twofold theory of pictorial seeing also owes to Wollheim (1980, 1998). For Wollheim, central to pictorial seeing,

> Is a special perceptual skill, called "seeing-in," which we, and perhaps the members of some other species, possess. Seeing-in is prior, both logically and historically, to representation. Logically, in that we can see things in surfaces that neither are nor are taken by us to be representations, say, a torso in a cloud, or a boy carrying a mysterious box in a stained, urban wall. And historically, in that doubtless our remote ancestors did such things before they thought of decorating the caves they lived in with images of the animals they hunted.
>
> (Wollheim, 1998: 221).

Wollheim was particularly interested in artworks, and their capacity of their surfaces to be the objects of aesthetic experience, and to depict objects and events, something he calls "twofoldness" (1998: 221). Twofoldness refers to a single experience with "two aspects," namely "configurational" and "recognitional." The former is involved in seeing the configuration of the surface, and the latter in the recognition of the scene, objects, or figures depicted therein. According to the twofoldness of pictorial seeing, we simultaneously see the colours and shapes of the pictorial surface, and the apples these features depict.

Ignoring some of the complications with twofoldness, VR picturing fits easily within this account of pictorial experience, though of course technologically its fit is complicated by the intervening role of computers in the production of the pictorial surface. But fundamentally, the graphics program produces visual markers on the display screen(s) of the VR headset, and we quite spontaneously perceive these as configuring shapes, objects and spaces; and additionally, we recognise these elements are depictions of items such as apples. But the egocentric pictures of VR are also importantly different to most traditional modes of picturing because of the feelings of presence and possibilities of interaction they generate. Again, I have described at length elsewhere the technology that allows for these developments, but the key trick of VR picturing is to allow the VR viewer's *apparent perspective within the VR environment* to correspond to the *actual position of the picture viewer*, so that when the user moves their head or body, the VR display coordinates a corresponding change from the pictorial perspective. This trick is achieved by a combination of the interactive structure of a 3D picturing environment, the tracking of the user's movements and projection of these "into" the picturing space, and the stereopsis of a VR headset (Tavinor, 2021: 43–54).

This, incidentally, is where we can fill out our understanding of the status of virtual circles referred to earlier. Even though a virtual depiction of a circle might occasionally render a circle on a stereoscopic headset, given the movement of the perspective of the user in the egocentric VR graphical environment, usually it is unlikely to do so. Instead, what is produced on the display is an "occlusion shape," that is, a visual configuration, which in combination with a tracked perspective will be perceived as a circle (Hyman, 2006: 81). Usually, these occlusion shapes will not comprise circles but rather projected ellipses, but ultimately, because of perceptual constancy, their configurations will be perceived as circles within the environment. There is obviously a great deal more that could be said about spatial perception and VR—and, indeed, I judge that this pictorial account of VR will strongly draw on such considerations (Tavinor, 2021: 119–125)—but I think this is already enough to show that it is incorrect to think of virtual objects such as circles as objects remediated or reproduced in a virtual substrate, but rather, that virtual reality can be used to render pictorial surfaces which realise apparent circles and circular objects. But as we will find shortly, this is also not a matter of the circle being a *fictional circle*, because this additional conclusion can only be made once the use and intentions behind an instance of VR picturing are understood.

I'm not the first person to have considered the connection between virtual reality and picturing. While not referring to VR technology—she was writing in the 1950s—the philosopher of the arts Susan Langer proposed that pictures, and art more generally, generates a sense of virtual reality, and that "virtual space is the primary illusion of all plastic art" (1953: 72). While Langer is cited by Jaron Lanier as a source of the term, it is not entirely clear how influential her thoughts have been on thinking about virtual reality. Moreover, when she claims that virtual space is "only visual," and that "this space has no continuity with the space in which we live; it is limited by the frame..." Langer's views do not fully anticipate how modern VR, with its different modes of experience and interaction, and the seeming lack of a frame, might generate such a continuity, especially in augmented reality applications. Nevertheless, Langer's work does count as an important precedent for understanding pictures and other artworks as presenting virtual spaces.[5]

Indeed, the potential for VR and its depictive powers to contribute to art itself has long been noted. The art theorist Ernst Gombrich, having recently read an article in *Scientific American* on early VR flight simulators, concluded, "Maybe one day these technical developments will lead to the rise of a new art form, as did scene-painting in Ancient Greece and Brunelleschi's experiment in the early 15th century" (Gombrich, 1987).

VR Picturalism and Representation

VR picturalism still faces challenges. As noted, the framework will need to be expanded to encompass the other sensory and perceptual modes increasingly utilised in VR technology. However, once these challenges have been met, VR picturalism can be put to work resolving some of the apparent puzzles with VR. Particularly, it may help us with some of the metaphysical confusion surrounding the question of precisely *what* it is we encounter in VR environments. Are VR objects and environments fictional or real?

For the VR picturalist, this question comprises a false dilemma. As a medium, VR may clearly be employed to represent both the fictional and the real. Many of the objects of VR depictions are fictional: when a headcrab lurches toward me in the VR game *Half-Life: Alyx*, I see the features of the pictorial surface—blocks of colour, shapes, apparent movement—I see the three-dimensional configurations apparent in these visual markers, and I fictionally see the head crab.[6] A stereoscopic headset may also be used to depict to its user real documentary objects: *Google Earth VR* depicts very real geographical features, though in an egocentric pictorial manner, but beyond this media difference there is no reason to think the experience is anything other than the viewing of pictures of real places. VR may even depict objects that are real and present to a user, and which they may really interact. My favourite example here is Mathew Pan and Günter Niemeyer's ball catching experiments at Disney Labs (2017), where a VR user may catch a ball depicted to them in a rudimentary VR environment, but there are many other such cases where the technology can be used to represent features of the real world with which the user can interact.

In fictional and realistic uses of VR the medium employed is the same, because each involves a dynamic and interactive pictorial display, a stereoscopic headset, and the tracking of the user's actual perspective into the apparent depictive space. What is different between the cases is the origin of the pictorial image, and the use to which it is put. *Half-Life's* headcrab is the product of an art department and an algorithm that displays it so that the user can be taken up in the fiction of battling interdimensional creatures in a hostile environment; the images in Google Earth VR involve satellite imagery and programs that augment this documentary visual content so that it can be viewed in an egocentric VR setting, and so that the user can learn about the actual world; in the Disney ball-catching experiment the real and present objects that are viewed and interacted with are tracked directly into the rudimentary VR environment.

So, what is it we see and encounter in VR? The answer to this will differ, depending both on the focus of our interest in VR pictorial experiences, and the use to which it is being put on a given occasion. First, VR users always see and encounter the screens and images before their eyes, and, if they are

to correctly perceive the pictorial content, they see the apparent spatial configurations perceptually inherent in these images.[7] But beyond these basic
acts of pictorial viewing, the user may see that there is a headcrab lurking
in the corridor; see the chaotic layout of the streets of Lower Manhattan;
or see the ball that is travelling toward them. It would be a mistake to col-
lapse this variety of objects into the view of there is some kind of *digital or
virtual object* that is what is seen in all cases, as is tempting to the virtual
digitalist, because this would be to conflate the basic features of pictorial
seeing, which are shared between the various cases, with the intentional object of the pictures, which is not. Of course, in each of the examples above,
the VR system involves a computational artefact that encodes the spatial
information that is used to generate the image on the screen, but this is not
what the user sees, as it does not even have the visual properties—shapes
and colours—that comprise the content of visual experiences. To conclude
so would be akin to the mistaken inference that in ordinary vision, one sees
the activity of neurons in the optic tract of the brain.

By eliminating these perceptible virtual objects, VR picturalism does
much to clarify the nature of VR media, and we might imagine that had
VR been understood in terms of its mode of picturing from the beginning, there might have been less urgency to ontologically problematize
the technology. VR is not the first time that technology has delivered us
with a new representational artifice, and with it, a new set of philosophical and theoretical puzzles. It is hard to think though, of cases where the
new technology has been theoretically framed in terms of *what* it represents, rather than *how* it does so. Most famously perhaps, photography
has been considered as an art of "mechanical" visual reproduction, and
while this mode of image production does give rise to certain puzzles related to the representational content thus produced, such as its potential
as art (Scruton, 1983), the medium itself can be understood via theoretical
means developed to assess the qualities of earlier modes of image production. Photography has also sometimes been claimed to have certain *epistemic* virtues, perhaps reproducing reality in a "transparent" way that is
impossible for previous modes of image making (Walton, 1979). But this
is an understanding born out of a close examination of the features of its
medium, its mechanical reproduction of the visual features of reality, and
its potential contribution to the practices of documentary and fiction making. This, I have argued, is not what has happened with the philosophical
understanding of VR technology, where the images have taken on an ontological life of their own. It would have been exceedingly odd if the inven-
tion of photography had initiated as its central concern the metaphysics of
a putative "photographic reality," and that a large philosophical literature
then sprang up devoted to explaining its ontology and value. This, however, is the curious case with VR.

Notes

1 Several papers in a recent issue of *Disputatio* (9 [46]), devoted to VR, take just this starting point.
2 Chalmers' *Reality+* (2022), a book that has a great deal of popular attention seems best read in this spirit.
3 The apples in VR *Minecraft* are quite stylised, and so even though they are VR, they are not particularly realistic.
4 This account obviously biases toward the visual aspects of VR, but I would also argue that the conceptual framework can be adapted to explain other perceptual and interactive modes, though I will not do so here.
5 More recently, the philosopher of the arts John Dilworth has taken a direct interest in virtual reality and perception, in a way that is largely consistent with the ideas presented here (Dilworth, 2010).
6 How to properly characterise acts of "fictional seeing" is of course a residual issue with the VR pictorialist account, but it is a problem that is shared with fictive media more generally.
7 Users need not pay any attention or even notice this screen, of course, and the designers of VR have put a great deal of effort into making the pictorial surface less prominent; but as seen earlier, this is also the case with traditional forms of depiction such as *trompe l'oeil*.

References

Anderson, p. L., Price, M., Edwards, S. M., Obasaju, M. A., Schmertz, S. K., Zimand, E., & Calamaras, M. R. (2013). Virtual reality exposure therapy for social anxiety disorder: A randomized controlled trial. *Journal of Consulting and Clinical Psychology*, 81 (5): 751–760.

Aulisio, M. C., Han, D. Y., & Glueck, A. C. (2020). Virtual reality gaming as a neurorehabilitation tool for brain injuries in adults: A systematic review. *Brain Injury*, 34 (10): 1322–1330.

Baudrillard, Jean. (1994). *Simulacra and Simulation*. Ann Arbor: The University of Michigan Press.

Bostrom, N. (2003). Are we living in a computer simulation? *The Philosophical Quarterly,* 53 (211): 243–255.

Brey, p. (2014). The physical and social reality of virtual worlds. In *The Oxford Handbook of Virtuality*, M. Grimshaw (ed). Oxford: Oxford University Press.

Castronova, E. (2005). *Synthetic Worlds*. Chicago: University of Chicago Press.

Chalmers, D. J. (2003). The matrix as metaphysics. In *Philosophers Explore the Matrix*, C. Grau (ed). Oxford: Oxford University Press.

Chalmers, D. J. (2017). The virtual and the real. *Disputatio,* 9 (46): 309–352.

Chalmers, D. J. (2022). *Reality+: Virtual Worlds and the Problems of Philosophy*. London: Allen Lane.

Conn, C., Lanier, J., Minsky, M., Fisher, S., & A. Druin. (1989). Virtual environments and interactivity: Windows to the future. *ACM SIGGRAPH 89 Panel Proceedings*, 7–18: New York ACM.

Dilworth, J. (2010). Realistic virtual reality and perception. *Philosophical Psychology,* 2: 23–42.

Dockx, K., Bekkers, E. M. J., Van den Bergh, V., Ginis, P., Rochester, L., Hausdorff, J. M., Mirelman, A., & Nieuwboer, A. (2016). Virtual reality for rehabilitation in Parkinson's disease. *Cochrane Database of Systematic Reviews,* 12: 1–50.

Feagin, S. (1998). Presentation and representation. *The Journal of Aesthetics and Art Criticism,* 56 (3): 234–240.

Gibson, W. (1984). *Neuromancer.* New York: Ace.

Gigante, M. A. (1993). Virtual reality: Definitions, history and applications. In *Virtual Reality Systems,* R. A. Earnshaw, M. A. Gigante & H. Jones (eds). London: Academic Press: 3–14.

Goodman, N. (1976). *Languages of Art: An Approach to a Theory of Symbols.* 2nd ed. Indianapolis, Ind: Hackett.

Goodman, N. (1979). *Ways of Worldmaking.* Indianapolis: Hackett.

Gombrich, E. H. (1987). Western art and the perception of space. *Space in European Art.* Japan: Council of Europe Exhibition: 5–12. Available at https://gombricharchive.files.wordpress.com/2011/04/showdis25.pdf

Grabarczyk, P., & Pokropski, M. (2016). Perception of affordances and experience of presence in virtual reality. *AVANT. Trends in Interdisciplinary Studies,* 25–44.

Gualeni, S., & Vella, D. (2020). *Virtual Existentialism: Meaning and Subjectivity in Virtual Worlds.* London: Palgrave Pivot.

Heim, M. (1993). *The Metaphysics of Virtual Reality.* Oxford: Oxford University Press.

Heim, M. (1998). *Virtual Realism.* Oxford: Oxford University Press.

Hyman, J. (2006). *The Objective Eye: Color, Form, and Reality in the Theory of Art.* Chicago: University of Chicago Press.

Jiang, M. Y. W., Upton, E., & Newby, J. M. (2020). A randomised wait-list controlled pilot of one-session virtual reality exposure therapy for blood-injection-injury phobias. *Journal of Affective Disorders,* 276: 636–645.

Kardong-Edgren, S., L. Farra, S., Alinier, G., & Michael Young, H. (2019). A call to unify definitions of virtual reality. *Clinical Simulation in Nursing,* 31: 28–34.

Krueger, M. (1983). *Artificial Reality.* Boston: Addison-Wesley.

Langer, S. (1953). *Feeling and Form: A Theory of Art Developed from Philosophy in a New Key.* New York: Charles Scribner's Sons.

Laver, K. E., Lange, B., George, S., Deutsch, J. E., Saposnik, G., & Crotty, M. (2017). Virtual reality for stroke rehabilitation. *Cochrane Database of Systematic Reviews,* 11: 1–161.

Ludlow, p. (2017). Cypher's choices: The variety and reality of virtual experiences. In *Experience Machines: The Philosophy of Virtual Worlds,* M. Silcox (ed). London: Rowman and Littlefield.

Milgram, P., & Kishino, F. (1994). A taxonomy of mixed reality visual displays. *IEICE Transactions on Information and Systems,* 77: 1321–1329.

Nozick, R. (1974). *Anarchy, State and Utopia.* New York: Basic Books.

Pan, M. K. X. J., & Niemeyer, G. (2017). Catching a real ball in virtual reality. *IEEE Virtual Reality* (VR): 269–270. https://ieeexplore.ieee.org/xpl/conhome/7889401/proceeding

Skarbez, R., Smith, M., & Whitton, M.C. (2021.) Revisiting Milgram and Kishino's reality-virtuality continuum. *Frontiers of Virtual Reality,* 2: 64.

Scruton, R. (1983.) Photography and representation." In Roger Scruton (ed), *The Aesthetic Understanding*. London: Methuen.

Tavinor, G. (2014). Art and aesthetics. In *The Routledge Companion to Video Game Studies*, M. J. p. Wolf & B. Perron (eds). New York: Routledge.

Tavinor, Grant. (2021). *The Aesthetics of Virtual Reality*. New York: Routledge.

Walton, Kendall (1979). Style and the Products and Processes of Art. In Leonard B. Meyer & Berel Lang (eds.), *The Concept of Style*. Philadelphia: University of Pennsylvania Press: 45–66.

Wollheim, R. (1980) " Seeing-as, Seeing-in, and Pictorial Representation." In R. Wollheim (ed.), Art and Its Objects: With Six Supplementary Essays. 2nd ed., pp. 205–226. Cambridge: Cambridge University Press.

Wollheim, R. (1998). On pictorial representation. *The Journal of Aesthetics and Art Criticism*, 56 (3): 217–226.

4 The Intersecting Frontiers of Extended Reality and Neuropsychology

Thomas D. Parsons and Joseph Neisser

Moving Neuropsychological Assessments Out of Flatland

Much of neuropsychology's history includes findings from well-controlled laboratory studies using low-dimensional stimulus presentations and basic technologies (i.e., static stimuli; limited interactivity; and text-based vignettes). Important knowledge about human nature has resulted from these low-dimensional stimulus presentations. There are times, however, when the parsimony offered by low-dimensional stimulus presentations may fall short of conveying the much higher-dimensional phenomena found in social, affective, and cognitive constructs. In fact, low-dimensional stimulus presentations may at times offer diminished interpretations of complex phenomena.

Some have compared the low-dimensional stimulus presentations in psychological science in general (Jolly & Chang, 2019) and neuropsychology specifically (Parsons & Duffield, 2020; Parsons, Gaggioli, & Riva, 2020) to the Flatland perspective. In Edwin Abbott's (1952) Flatland text on perception and dimensionality, the Flatlander A. Square (Abbott's narrator) is only capable of perceiving two dimensions. A. Square encounters a "Stranger" (a sphere) who guides Square's perspective into an understanding of the actual complexity (higher dimensionality) of the world. Likewise, in neuropsychology, low-dimensional stimulus presentations may result in simplified explanations of complex phenomena, which may in turn limit the usefulness of models of human cognition, affect, and social interactions. Jolly and Chang (2019) call for psychologists to move beyond this "Flatland fallacy" via formalizations of psychological theories as computational models that can produce detailed neurocognitive predictions (see Figure 4.1).

The Flatlander's constrained and low-dimensional perspective (bottom of the figure) leads Square to perceive the three-dimensional sphere as a circle of varying sizes (increasing and decreasing radii). From the top of the figure, we can see that the object is a sphere that is progressing across a

DOI: 10.4324/9781003359494-6

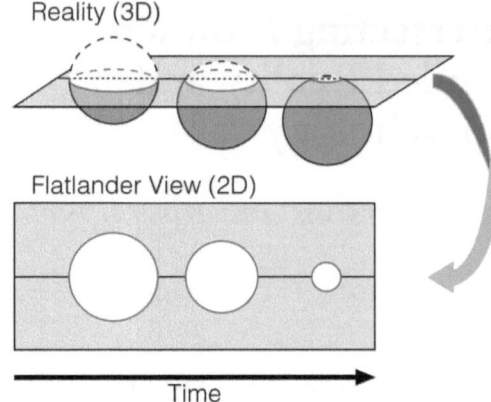

Figure 4.1 A square cannot perceive his world as anything other than two-dimensional. (Figure adapted from Jolly and Chang (2019) (Open Access; Creative Commons).

lower-dimensional plane. The Flatland (low-dimensional) perspective limits the Flatlander's perception and understanding of reality. Comparably, psychologists may at times fuzzily conclude that perceptions from a low level of dimensionality comprehensively explain cognitive, affective, and social phenomena. The Flatland narrative calls attention to the need for the development of models from high-dimensional stimulus presentations that better reflect the reciprocal relations among persons interacting with others in various environments (Beauchamp, 2017).

In the human neurosciences (clinical, social, and cognitive), there is increasing interest in an enhanced understanding of complex and dynamic interactions involved in the brain's processes (Kennedy & Adolphs, 2012; Parsons, 2015; Wilms et al., 2010). Moreover, there is mounting interest in the development of high-dimensional tools for assessing and modeling brain functions using high-dimensional presentations of environmentally relevant stimuli (see Kane & Parsons, 2017 for a book-length review; see also Parsons & Duffield, 2019; Zaki & Ochsner, 2009).

Technological enhancements that include high-dimensional stimulus presentations hold promise for enhancing psychological science. While psychologists have long used high-dimensional assessment technologies like functional neuroimaging for real-time observations of brain functioning, less work has been done using high-dimensional tools such as those found in extended reality (e.g., virtual/augmented reality) platforms that offer enhanced stimulus presentations and interactivity. Extended reality (XR) is a collective term for immersive technologies, including (1) virtual reality (VR) which fully immerses the participant into a computer-generated virtual world; (2) augmented reality (AR) which overlays digitized content onto the real world; and (3) mixed reality (MR) which includes virtual objects that are not just overlaid on the real world but can interact with it. In the same way that technological innovations are constantly evolving, the definitions of XR are consistently developing (Palmas & Klinker, 2020). Psychologists observing people as they interact in extended reality environments may redefine earlier understandings of social, cognitive, affective, and behavioral functions (Price, 2018). That said, the use of extended reality environments in psychology has not found the same level of adoption as neuroimaging. Nevertheless, extended reality platforms can offer some advantages. For example, extended reality allows for the design and use of simulations reflecting everyday activities and interactions.

Balancing Ecological Validity with Experimental Control

The naturalistic and high-dimensional stimuli found in extended reality platforms also offer enhanced control of the various extraneous variables that can be problematic (even impossible) for experimental control in naturalistic observation studies emphasizing ecological validity (Dombeck & Reiser, 2012; Sauzéon et al., 2012). Moving beyond the low experimental control found in ecologically valid naturalistic observations, extended reality environments balance high-dimensional stimuli with experimental control. Hence, extended reality platforms offer promise for high-dimensional stimulus presentations of real-life activities and interactions (i.e., ecological validity) that also maintain experimental control (Bohil, Alicea, & Biocca, 2011; Parsons, 2015; Parsons, Gaggioli, & Riva, 2017).

Moreover, low-dimensional stimulus presentations are often limited to abstract cognitive constructs. Neuropsychologists using these measures attempt to generalize findings from low-dimensional stimuli to the complexities found in everyday cognitive functions. Burgess and colleagues (2006) argue that this attempt at generalization is going in the wrong direction. Their proposed alternative is that neuropsychological assessments should involve tasks that reflect real-world "functions" and move from directly observable everyday functions (e.g., behaviors) to the ways in which action sequences lead to given cognitive functions. Extended reality platforms offer high dimensional and function-led measures for psychological assessments (Parsons et al., 2017).

Importantly, there is reason to suspect that the results obtained by using XR tools are significantly different from those obtained using less ecologically realistic and/or less well-controlled techniques. Because there is increasing speculation about how much confidence we can have about historical results found in psychological science studies (Nosek & Errington, 2020; Nosek et al., 2022; Open Science Collaboration, 2015;

Pashler & Wagenmakers, 2012), psychological science must demonstrate that its findings are robust and replicable. This is an area that virtual reality can assist. Take for example the typical finding that in studies using low-dimensional stimulus presentations have found that males outperform females on 2D mental rotation tasks (Voyer, Voyer, & Bryden, 1995). In fact, a quantitative meta-analysis from Voyer and colleagues (1995) found that the performance difference between males and females on a 2D mental rotation task revealed an average effect size of 0.94. From this, it has been surmised that males on average outperform females by almost a full standard deviation. This viewpoint has long been accepted. However, other reviews of gender differences in cognitive abilities call into question the effect size and conclusions of these studies (Hyde, 2005). What happens when we move to high-dimensional 3D presentations? Researchers have found no significant gender effect when mental rotation tasks are conducted using high-dimensional 3D models (Monahan et al., 2008; Parsons et al., 2004). In terms of replicability, this raises an issue that the commonly reported sex effect really is an artifact of traditional laboratory conditions. (Lochhead et al., 2022).

The limitations of sterile laboratory findings to the processes normally occurring in people's everyday lives have long been discussed in the assessment of memory (Neisser, 1978). While the presentation of everyday objects as low-dimensional pictorial stimuli on a computer monitor allows for rigorous control of the stimuli's features, their actual meaningfulness is obscured as cortical object processing does not involve processes entirely invariant to an object's physical properties (Holler et al., 2020). Memories found in conventional laboratory events inadequately reflect engrams formed under realistic (complex) conditions (Cabeza et al., 2004; Cabeza and St Jacques, 2007). Results from virtual reality studies reveal successful facilitation of the formation of profound memory traces (Kisker et al., 2021a, 2021b; Schöne et al., 2019). This suggests that virtual reality immersion weaves experiences into a participant's narrative or autobiographical memory in a manner similar to everyday activities. Thus, virtual reality platforms may offer ways to enhance research into the neural correlates of encoding for manipulations that remain inappropriate for testing in a real-life context (e.g., Bréchet et al., 2019; Vass et al., 2016).

Concerns about High-Fidelity Stimulus Presentations

The high-dimensional presentation of visual, aural, tactile, and/or even olfactory information found in extended reality environments comes with philosophical and ethical concerns (Behr et al., 2005; Parsons, 2021). According to Slater and colleagues (2020), today's XR platforms represent a "super-realism" with greatly enhanced stereoscopic vision, head tracking, eye tracking, sound rendering, haptic rendering, and smell machines (olfaction). This super-realism in XR leads to questions about whether the effective recreation of real-world scenarios in VR results in ethical violations for research in human subjects (Ramirez, 2019; Ramirez & LaBarge, 2018, 2020).

In addition to the enhanced stimulus presentation found in these super realistic XR platforms, adaptive algorithms, and wearable sensors may extend users' cognitive, affective, and social processes beyond the wetware of their brains. This is apparent in technologies ranging from smartphones that enable us to navigate, translate, recall, analyze, and compute information to adaptive virtual environments that use machine learning to personalize the user's experience. While much of the work in XR has focused on stimulus presentations to individuals, the advent of social VR and the Internet of Things (IoT) connects everyday objects (including virtual simulations) to the Internet and enables data transfer from network-connected devices to remote locations. The rise of telepsychology and social virtual reality calls for increased attention to a Virtual Environment of Things (VEoT) that extends the user's experience of real-world smart technologies with virtual objects and avatars in interactive and adaptive virtual environments (Lv, 2020; Wu, Chou, & Jiang, 2014).

Increasingly, much of this information is publicly available via the Internet. Smart technologies using active and passive data analytics gather personal information (e.g., contacts, emails, text messages, posts, and calendar appointments) and log everyday activities (purchases, readings, film viewing, steps taken, calories, and so forth). These smart technologies learn from their users, can be programmed to make suggestions, and extend the user's cognitive processes. Data needed for learning comes from participating public and private entities that collect and process data (including personal user information). For example, the Meta Quest Pro is a head-mounted display (HMD) that is used for virtual reality. The Meta Quest Pro logs eye tracking and facial expressions using five inward-facing cameras and can automatically upload this information to Meta servers. This technology extends the user's eye movements and facial expressions to the facial expressions (smiles, winks, or raised eyebrows) of the user's avatar in real time. While Meta does allow users to decide whether to share their personal information (e.g., biometrics like eye tracking and facial expressions) with the company, users are informed that such information is essential for the immersive experience. As a result, many users may share their data to better experience the virtual environments. It is important to note that the current lack of data governance for such platforms may result in user data being shared with third-party apps aimed at enhanced understanding and manipulations of the users' experience.

Extended Cognition, XR, and the Brain

The Parity Principle

The notion of extended cognition has slowly gained traction in philosophy and psychology. Daniel Dennett (1996) has stated that the human brain frequently extends our cognitive processes into the environment by

> Off-loading as much as possible of our cognitive tasks into the environment itself – extruding our minds (that is, our mental projects and activities) into the surrounding world, where a host of peripheral devices we construct can store, process and re-represent our meanings, streamlining, enhancing, and protecting the processes of transformation that are our thinking. This widespread practice of off-loading releases us from the limitations of our animal brains.
>
> (pp. 134–135)

One may view cognitive processes as being performed by both our brains and the technologies we use (Clark & Chalmers, 1998). According to this extended mind perspective, human cognition is processed via a system coupled with the environment (Clark, 2008). Clark and Chalmers (1998) propose the "parity principle" for analyzing the extension of extended cognitive systems from brain-based cognitive processes to external resources (e.g., technologies such as smartphones and the Internet):

> If, as we confront some task, a part of the world functions as a process which, were it to go on in the head, we would have no hesitation in recognizing as part of the cognitive process, then that part of the world is (so we claim) part of the cognitive process.
>
> (Clark & Chalmers, 1998, p. 8)

To illustrate the parity principle, Clark and Chalmers employ fictional characters, Inga and Otto, who must navigate to the Museum of Modern Art (MOMA) on Fifty-Third Street in New York City. While Inga is able to recall the directions from her internal brain-based memory processes, Otto's Alzheimer's disease limits his ability to recall the directions from the sole use of his internal brain-based memory processes. Instead, Otto must use both his brain and an external aid found in a notebook. The notebook and brain are coupled in an information-processing loop that extends beyond the neural realm to include elements of our social and technological environments.

The cases of Otto and Inga illustrate that mental processes cannot be fully reduced to brain processes. Take, for example, the potential of smartphones connected to the Internet to extend our brain-based memory. Mobile technologies connected to the Internet allow for novel investigations into the interactions of people as they engage with a global workspace and connected knowledge bases. Moreover, mobile access to the Internet may allow for interactive possibilities: a shift in how we see ourselves and the ways in which we understand the nature of our cognitive and epistemic capabilities (Parsons, 2017).

Moving Beyond the Parity Principle: Technologies of the Extended Mind

The addition of algorithmic devices to the extended mind discussion suggests further changes to received conceptions of mental processing (Reiner & Nagel, 2017). According to Reiner and Nagel (2017), most of us are extending our cognitive processes via the algorithmic devices that we keep close at hand. Reiner and colleagues have termed these algorithmic devices as technologies of the extended mind, or TEM (Fitz & Reiner 2016; Nagel & Reiner, 2018; Reiner & Nagel, 2017). It is important to note that not every algorithmic function carried out by technologies external to the brain qualifies as a TEM. Instead, there must be a comparatively continuous interface between the brain and the algorithm such that the person experiences the algorithmic device as an extension of the person's mind:

> It is not the case that every algorithmic function carried out by devices external to the brain qualifies them as a TEM, but rather that there is a relatively seamless interaction between brain and algorithm such that a person perceives of the algorithm as being a bona fide extension of a person's mind. This raises the bar for inclusion into the category of algorithms that might be considered TEMs. It is also the case that algorithmic functions that do not qualify as TEMs today may do so at some future point in time and vice versa.
> (Reiner & Nagel, 2017, p. 110)

Reiner and Nagel illustrate this point using a new driver for Uber (i.e., a company that allows nonprofessionals to act as chauffeurs using their own automobiles) as an example. While the new Uber driver is offloading cognitive spatial navigation processing to a global positioning system (GPS) to navigate New York City, they argue that it is not yet a technology of the extended mind (TEM). It will not count as a TEM until the algorithmic calculations and the driver's reliance on them is seamlessly integrated with the driver's cognitive processes. This addition to Clark and Chalmers's parity principle specifies that the concept of an extended mind requires the presence of a relatively seamless interaction between the person's brain and the algorithm such that the person perceives the algorithm as a bona fide extension of their mind. We return to discuss this proposal in Section 3, below. First, we introduce some relevant neuropsychological considerations.

A Neuropsychological Framework for Understanding Technologies of the Extended Mind

Parsons (2019, 2021) proposed a framework for understanding technologies of the extended mind from a neuropsychological perspective. Parsons' tripartite approach conceptualizes a continuous functional interface between brains and algorithms in which persons both perceive (Reiner & Nagel, 2017) and "act as if" (Parsons, 2019, 2021) the algorithm is an actual extension of their mind. This approach builds on Stanovich's (2009) tripartite model of cognitive processing. In addition to an autonomous (automatic; rapid; and nonconscious use of heuristics) processor, Stanovich argued that controlled processing (slow; effortful) involves two subdivisions: (1) reflective processing characterizes the goals of cognitive processing, goal-relevant beliefs, and optimizing choices of action; and (2) algorithmic processing that includes "mindware" that consists of the rules, strategies, and procedures that a person can retrieve from memory to aid problem-solving.

A neuropsychologically informed version of Stanovich's tripartite model made up of three neural systems, has been proposed by Wood and Bechara (2014). The neuroscience of their triple system includes (1) automatic processing (fast, automatic, nonconscious, and habitual behaviors) via amygdalastriatal (limbic-ventral striatal loop structures such as ventral striatum and amygdala) system; (2) reflective processing (planning, prediction, and inhibitory control) via prefrontal-dorsal striatal loop (prefrontal cortex mediation of decision-making and inhibitory control); and (3) algorithmic interoception via the insular cortex. The limbic-ventral striatal loop (automatic) and prefrontal-dorsal striatal loop (reflective) processing can be assumed to act in parallel and can interact with each other during decision-making, whereby one system acts in a predominant role. Situational and environmental features influence the processes that activate a predominant system (Schiebener & Brand, 2015). The insular cortex offers a third (interoceptive awareness) system that activates representations of homeostatic states to translate somatic states into more conscious states (Noël et al., 2013) and modifies the equilibrium between the automatic and reflective system (Wood & Bechara, 2014).

Parsons (2019) extends the neuropsychologically-based tripartite approach to a tripartite processing model for technologies that extend cognition (see Figure 4.2). According to Parsons's framework, automatic algorithmic processes originating with an algorithmic device are coupled with the automatic (C-System: limbic-ventral striatal loop), reflective

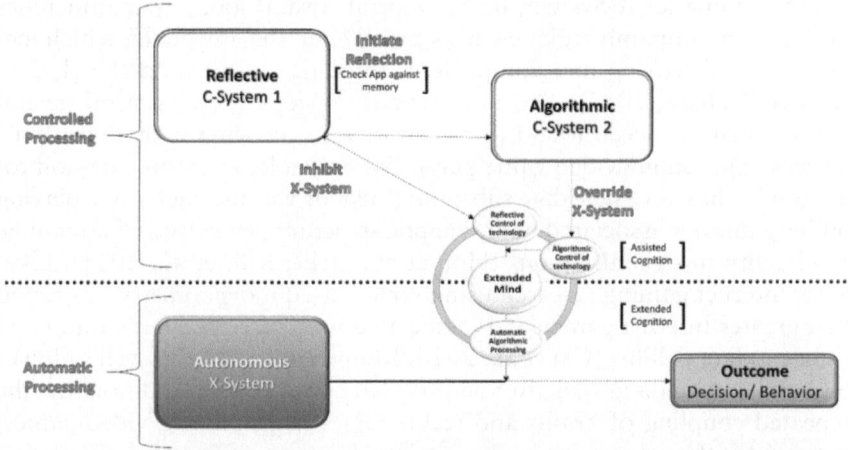

Figure 4.2 Framework for understanding technologies of the extended mind.

(C-System 1: prefrontal-dorsal striatal loop), and algorithmic (C-System 2: insular cortex) processing of the tripartite model. The algorithmic processes of technology can, over time, become an automated and algorithmic coupling of brain and technology. When the user first starts operating a new device, there is a period in which the user relies on controlled (C-System 1: reflective) cognitive processes found in the prefrontal-dorsal striatal loop to inhibit and override prompts initiated by the device (see reflective and algorithmic control of technology in Figure 4.2). After using the technology for a period of time, the algorithmic operations become overlearned and begin to rely automatically upon the limbic-ventral striatal loop (X-System: automatic processing). The extension of these brain processes to algorithmic technologies is balanced (C-System 2) by insula cortex processing of salient environmental factors that bias a technology user's deployment of automatic (X-System: limbic-ventral striatal loop) and reflective (C-System 1: prefrontal-dorsal striatal loop) information processing.

The use of algorithmic virtual environments and the Internet of Things (IoT) enables algorithmic coupling of users' brains and network-connected virtual environments. This Virtual Environment of Things (VEoT) extends the user's experience and allows the user to act as if virtual objects and avatars are real objects and agents. These VEoT platforms can manipulate users because the user experiences virtual rewards and penalties as if they were real. Despite being virtual, these reinforcement schemas are difficult to resist. Virtual environment and video game developers use these inducements to ensure that users (e.g., video gamers) keep using the algorithmic devices (e.g., virtual environments; video games).

The automatic (X-System: limbic-ventral striatal loop) system includes the striatum (dopaminergic reward systems) and the amygdala, which mediate reward-seeking and compulsion, through sensitization (Noël, Brevers, & Bechara, 2013). The automatic (X-System: limbic-ventral striatal loop) system has been found to be sensitive to coupling with algorithmic devices (e.g., online video game play). For example, positron emission tomography has revealed that substantial use of the Internet (e.g., playing online games) is associated with synaptic structure plasticity and dopamine availability in striatal regions (Hou et al., 2012; Kim et al., 2011). Likewise, Internet gaming research using voxel-based morphometry has found that greater Internet gameplay is associated with increased left striatal and right caudate volume (Cai et al., 2016; Kühn et al., 2011), as well as lower bilateral amygdala gray matter density (Ko et al., 2015). Additionally, the repeated coupling of brains and technologies (e.g., playing video games) strengthens the association between technology use and reward (Turel, Serenko, & Giles, 2011).

The reflective (C-System 1: prefrontal-dorsal striatal loop) system controls working memory and executive functions (e.g., inhibition of prepotent responses, mental set shifting). These controlled cognitive processes are primarily dependent on the prefrontal cortices and the anterior cingulate cortex. The algorithmic dependence that occurs in excessive video game play is associated with decreased functional connectivity in the prefrontal cortex (Jeong et al., 2016; Jin et al., 2016) and significant hyperactivity in the anterior cingulate cortex (Ding et al., 2014). Moreover, reduced fractional anisotropy in the dorsolateral prefrontal cortex and anterior cingulate cortex has been found in internet gaming disorder (IGD) (Yuan et al., 2016). These data indicate that excessive use of technologies impacts brain areas responsible for the critical abilities of the reflective system to suppress cognitive (Cai et al., 2016; Ko et al., 2014) and motor response inhibition (Ding et al., 2014). In addition to these neuroimaging studies that consistently report abnormalities in brain structure and function in internet gaming disorder, a number of quantitative meta-analyses have synthesized the literature (Niu et al., 2022; Solly et al., 2022; Sun et al., 2023) and revealed structural and functional impairments in brain regions related to executive cognitive control.

The algorithmic (C-System 2: interoceptive awareness) system relies on the insular cortex (insula). The insula is understood to be a gateway to visceral needs and mediates the generation of homeostatic perturbations (Craig, 2009; Zhao et al., 2023). The insular activity found in C-System 2 can stimulate motivation by biasing effective incentive inputs to feedback loops (Noël et al., 2013). The insula plays a role in excessive technology use (e.g., Internet gaming disorder) and neuroimaging studies have found decreased functional connectivity between the insula and the motor/executive cortices (prefrontal cortex; cingulate cortex) found in the reflective (C-System 1) system (Zhang et al., 2016). Furthermore, neuroimaging during videogame-related cues reveal evidence suggesting robust associations between the insula and the automatic (X-System) and reflective (C-System 1) systems (Ko et al., 2009).

To illustrate how Parsons's tripartite neuropsychological framework applies to TEMs, imagine a modification of the original case of Inga visiting the MOMA, introduced in Section 2.1 above. In this version, a telepresence robot allows Inga to virtually experience the museum "as if" she were actually there. The telepresence robot offers an all-in-one solution that includes everything Inga needs to virtually experience the Museum. Inga wears a head-mounted video display that allows her to see, hear, and smell. She also wears haptic gloves and a body suit that provides feedback, motion, and biometrics. Inga is able to navigate, interact, and move anywhere in the Museum utilizing a completely remote-controlled robot. The telepresence robot includes an integrated camera, microphones, and a motorized and remote-controlled platform. This allows Inga to freely move between exhibit locations, at Inga's own pace.

After being immersed in a Virtual Environment of Things (VEoT) simulation of the Museum, Inga can navigate using GPS coordinates of the real-world museum and a learning algorithm. Inga is informed that she can search for exhibits by entering them into the telepresence robot app that will show her the best route. Upon her arrival at a destination, the telepresence robot app can be used interactively by Inga to learn about the exhibit. Inga sees this as a benefit because it can keep her from getting lost when exhibits are in parts of the museum with which she was unfamiliar. In the beginning, Inga's lack of familiarity with this application results in some initial skepticism about the technology. As a result, Inga remains alert (see controlled/reflective processing of C-System 1 (prefrontal-dorsal striatal loop) in Figure 4.2) to her environment so that she can be sure that she makes it to the museum exhibits without a problem.

But though Inga is initially reliant on C-System 1 (prefrontal-dorsal striatal loop) and vigilantly double-checked the feedback from the telepresence robot Museum application, she soon comes to simply accept the telepresence robot app most of the time, and only rarely does she stop herself from automatically (X-System: limbic-ventral striatal loop) following the application's guidance (see Figure 4.2). Here one may ask: Are Inga's telepresence robot app and the GPS functioning as a technology of the extended mind? While it is undoubtedly performing computations that are external to Inga's brain, Parsons argues that the functions of the GPS in Inga's telepresence robot app are best understood as cognitive assistance rather than an extension. Why is this the case? The answer is that neither the algorithmic calculations from the telepresence robot app nor Inga's use of them are automated with Inga's cognitive processes (see algorithmic control of technology in Figure 4.2).

Now consider a different scenario in which Inga has experienced the exhibits several times over the course of a month. Even though she now has slightly more knowledge of the museum, she consistently uses the GPS in her telepresence robot app to navigate through the museum. The telepresence robot app has not failed her in its directions to exhibits or its information (e.g., artist, history, and subtleties of the work) about the art at each exhibit. At this point, when she enters an exhibit into the telepresence robot app's search interface and the route is presented on the screen of the head-mounted display, she automatically (X-System: limbic-ventral striatal loop) follows it to the destination suggested by her telepresence robot app and readily receives information about the art. The telepresence robot app is beginning to function as a technology of the extended mind because Inga has integrated its algorithmic processes into the working of her mind.

From Extended Cognition to Extended Selves

Two Kinds of Extended Cognition

The parity principle can be used to generate several different versions of the extended cognition thesis. In a later paper, Chalmers (2019) distinguishes two very different kinds of extension: *circuit extension* vs *sensorimotor extension*. Consider these in turn.

Circuit extension is exemplified by the case of Diva, a human who suffers minor brain damage and loses some specific cognitive ability, say, arithmetic (Clark, 2009)). Diva has an external silicon circuit connected wirelessly to her brain, restoring the original functions without loss. The parity principle is then invoked, and the conclusion is that Diva's cognition is extended across (constituted by) the brain-computer system, not just the brain. This conclusion seems almost undeniable –at least no functionalist should deny it. In fact, even the opponents of extended cognition do not deny it (Adams & Aizawa, 2001).). And this is also the problem with it –it is too weak to generate a distinctive thesis. In particular, it fails to give guidance for our question here: when does a distributed brain-body-world cognitive system constitute an extended *self* or *person*?

Sensorimotor extension, on the other hand, is different. Here, cognition is extended via perception and action, rather than through direct circuitry. Perceptual systems become modules within a larger brain-body-world cognitive architecture. To see this, return to the original examples from Clark and Chalmers. Otto accesses the notebook by *reading* it, while the Tetris player performs epistemic manipulations by *rotating* the shapes to decide where they fit. Invoking the parity principle in these cases yields the conclusion that reading the notebook is a case of remembering and that rotating the shapes is a case of mental rotation. Chalmers (2019) argues that this element of "interacting via perception and action" is what makes the thesis of extended cognition interesting and distinctive, and what renders the mere circuit extension, displayed in the case of Diva, trivial. (p. 6).

Initially, fans of dynamic systems theory and cybernetics might think this is a distinction without a difference. In both cases, they might think, the external resource is coupled with the brain in a way that constitutes a distributed cognitive system. But Chalmers argues that the distinction is important because the latter, sensorimotor version of the extended cognition thesis indicates that one's cognitive processes can be outside of *oneself* in surprising ways. Chalmers eventually settles on this revision of the extended mind thesis:

> A subject's cognitive processes and mental states can be partly constituted by entities that are external to the subject, in virtue of the subject's sensorimotor interaction with these entities.
>
> (Chalmers, 2019, p. 7)

For the present purpose, the key line is that cognitive processes can be external, not just to the skin and skull, but *to the subject whose cognitions they are*. This kind of case is importantly different from the Diva case. There, the extensions are entirely subpersonal. But epistemic actions such as rotating the Tetris shape or looking up the address partly depend on person-level processes that are paradigmatically conscious. (It is a further question whether unconscious or unattended sensorimotor interactions could qualify, though there seems no principled reason why not.) Chalmers explains that the reason that sensorimotor extension is controversial, while circuit extension is not, is that it violates a deep set of assumptions in psychology, namely, that the cognitive domain is bounded by perception and action – i.e., that thought is what lies *between* perception and action. Indeed, this idea is fundamental to cognitive psychology. Thus, if the thesis of sensorimotor extension is correct, then a founding assumption of cognitive psychology is apparently false. In contrast, the extended circuit thesis is just another piece of functionalism. As Chalmers puts it, "The thesis does not just overthrow the hegemony of skin and skull as boundaries for cognition (as we claimed at the end of the original article). It also overthrows the hegemony of perception and action." (p. 9)

There are several implications of this line of reasoning for the issue of whether – and when –an extended cognitive system can constitute an extended self. As it stands, the distinction between circuit extension and sensorimotor extension suggests a simple way of proceeding. Circuit extensions would be part of the extended self because they are functionally "between" the inputs and outputs –perception and action – that constitute the bounds of an experiencing subject. So far so good; let it be agreed that in the not-so-distant future when subpersonal brain-computer interface (BCI) technologies gain wider clinical application, cases such as Diva will constitute *extended selves*. As such, a range of ethical issues will be immediately in play. Klein (2017) addressed several of these issues.

By this criterion, however, sensorimotor extensions of the kind Chalmers champions would *not* be part of the extended self since they would not be between input and output. One must interact with them, as it were, "personally," hence they are not part of one's person. But notice that Chalmers does not speak of the self or the person directly. Rather, he speaks of the *subject*. Subjectivity marks a connection to consciousness, and it is to the possibility of extended consciousness that Chalmers then turns. He argues that perception and action still constitute the bounds of consciousness, even when cognition transcends these bounds via sensorimotor extension. Paradigmatically conscious states are correlated with "direct availability for global control" (p. 11), a criterion that is broadly in tune with Global Neuronal Workspace Theory (Baars, 1988; Dehaene, 2014), with analyses of *access consciousness* Block (1995), or with the representational account in Tye (1995). Roughly, Chalmers holds that information accessible and manipulable only through perception and action is not directly available for global control, even though it may be part of an extended cognitive system. Through perception, one can become conscious *of* this information, but the information process per se is not a correlate of consciousness.

A Third Kind of Extended Cognition? Extensions, Embodied Selves, and Neural Mechanisms

But what of the self? Must we stay home, as it were, with consciousness? Or can the self, along with cognition, "overthrow the hegemony of perception and action"? In the philosophy of mind, the term "subjectivity" is used to refer to a particular feature of consciousness, namely, its first-person aspect (e.g., Kriegel, 2009; Levine, 2001; Neisser, 2015). But in the wider literature, the self is rarely identified with anything fully conscious or fully explicit. And it seems intuitively clear that various nonconscious dimensions of embodiment, including many unconscious or subpersonal processes "within" the mind-brain, would count as partly constitutive of the self – this is what the case of Diva establishes. Thus, the self can extend beyond consciousness in much the way that cognition can. By parity, then, the way seems open to sensorimotor extensions of the self, too. It seems possible that the self might venture forth beyond the bounds of the conscious subject.

In the multidisciplinary literature on technology studies, neurophilosophy, and posthumanism, there are many, many attempts at "rethinking"
the self that would be broadly consonant with this idea. Three that are
relevant to our present purpose are Gallagher (2013), "A pattern theory of
the self," Hayles (2017) *Unthought: The power of the cognitive nonconscious*, and especially Seth (2019) "Being a beast machine."

The pattern theory is a pluralistic approach to the self, inspired by
so-called pattern theories of emotion (e.g., Mendoça, 2012). Gallagher
(2013) suggests that "the self" is a cluster concept, and that selves can
be constituted by multiple and partially overlapping assemblages ("patterns") of characteristic elements. Gallagher suggests that, typically, selves
are embodied, experiential, affective, cognitive, intersubjective, narrative,
extended, and situated (p. 4). Gallagher remains officially agnostic about
whether any of these elements might be individually necessary, but he is
clear that no subset of these elements is both necessary and sufficient. Different self-assemblages are dynamic gestalts that mix and match these (and
possibly other) elements in different ways, forming a set of family resem-
blances. So the pattern theory seems to allow for selves that might over-
flow the bounds of perception and action that constrain consciousness.

But Gallagher's treatment of *extended* self-patterns is undeveloped. Here
is his sole comment on the extended aspect of patterned selves:

> James (1890) suggested that what we call self may include physical
> pieces of property, such as clothes, homes, and various things that we
> own. We identify ourselves with stuff we own, and perhaps with the
> technologies we use, the institutions we work in, or the nation states
> that we inhabit.
>
> (p. 4)

For our purpose here, this is inadequate. At a minimum, an account of the
extended self should distinguish between what we value and what we are.
That I value something or "identify with it" cannot be sufficient to make it
literally a part of me, at least not in the sense that is of interest in the debate
about technologies of the extended self. There is something different about
the ways we become entwined and entangled with technology. Our cyber-
entanglements don't just arise because we *really like* our technology, or
because technology is important for work and home life. So the idea that
extensions can become part of a self-pattern needs something more than
what Gallagher and James have offered.

One simple possibility here would be to hold that there is a self wherever
there is cognition of any kind, i.e., that the categories of "cognition" and
"self" are coextensive. In that case, any cognitive process, by definition,
constitutes a self. But then, what of stand-alone computational devices and

systems (or what today is all too easily referred to as "AI")? While these might well be classified as cognitive systems, it would be odd to assume, on that basis alone, that they are (or have) selves. It seems clear that there may be information processes without a self. In addition, this move would seem to simply ignore Chalmers' distinction between circuit extension and sensorimotor extension, and his argument that the interest and power of the extended cognition thesis is precisely to distinguish cognition from self or subject, such that the former *can* at times extend beyond the latter. We don't want to ignore this distinction but rather to develop it in a way that sheds light on the relationship between cognition, consciousness, and the self.

Another possibility might be that whenever a *previously existing* self becomes cognitively extended, then the extensions are part of that self. This, at least, partly captures the idea that my thoughts, being mine, are part of me. If the information processes in the extended system are said to be "my thoughts," then this may be a prima facie reason to think that the cognitive extensions in which they take place are part of "me." But this, again, ignores the distinction between circuit and sensorimotor extension, fails to shed any new light on what a self is, and remains inadequate to an understanding of our "entanglement" with technology.

Whatever an extended self is, it is not just an extension of thought but of *embodiment*. A self is not just a "thinking thing" (though it is that). It is also a particular kind of being in the world, a way of being that distinguishes itself from its world, or carves itself out from the world. So, an extended self is more than extended cognition and more than an extended body. It is a self-extension that is somehow *lived*. Just as the self is not merely whatever one thinks about or values, the self is also not just whatever bodily substrate realizes one's cognition.

If that is right, then the parity principle that grounds the attribution of extended cognition cannot be sufficient, by itself, to ground the attribution of an extended self. What further kinds of processes could ground self-extension, then? Just as pressing: when extended selves do arise, how might they enact kinds of being in the world that are genuinely "post-Cartesian"? How can an investigation of the extended self also contribute to "rethinking" received ideas in the humanistic tradition? When the prototypical human subject is just one possible formation of the self, one possible pattern for a self, the effect is to "decenter the human" and to suggest a continuum between human, animal, and technological forms of cognition (Hayles, 2017, p. 182). Here, the self becomes a crucial category that is necessary for retaining contact with, while simultaneously transforming, a range of moral and practical concepts like agency, responsibility, integrity, and privacy. In the post-human world of technological entanglement and extended embodiment, these categories will not simply be jettisoned. Instead, they must be retrofitted for new applications.

Predictive Coding and the Extended Self

The way forward for an account of the extended self is through the brain. It is worth recalling that one of Andy Clark's initial motivations for the extended cognition thesis was his commitment to a kind of embodied cognitive neuroscience in which neural mechanisms are conceived as cybernetic controllers or governors rather than as good old-fashioned von Neuman machines (e.g., Clark, 2008). Despite its commitment to embodiment, Clark's research program was by no means "anti-cognitivist." Instead, it espoused minimal cognitivism in which the content of cognition supervenes on a distributed system that includes but is not limited to neural mechanisms. This program has been extremely successful, ultimately producing (or at least converging on) the concept of *predictive coding* as a central explanatory idea in cognitive neuroscience (Clark, 2015; Friston, 2010; Hohwy, 2013).

Predictive coding models of cognition posit an unconscious Bayesian process of active inference, in which sensory inputs are used to generate an expectation about the next inputs. In turn, these Bayesian posteriors constitute an inference about the causes of stimuli. Anil Seth (2019) speculates that this framework might also help explain selfhood. He argues that feelings of selfhood are the result of a certain kind of predictive processing – *control-oriented interoceptive inference*. Seth distinguishes two varieties of feedback mechanisms: *epistemic* and *control-oriented*. In the former, actions are taken in order to refine model-based representations of the causes of sensory signals. These epistemic predictive processes are primarily employed in the perception of the external world. In contrast, control-oriented active inferences involve actions taken to regulate feedback and to keep it within a homeostatic range (pp. 242–243). Seth argues that interoception (as opposed to perception and proprioception) is a matter of control-oriented predictive coding for allostatic regulation (p. 245). Further, he believes that the distinction between epistemic and control-oriented cybernetic mechanisms may account for the difference between the perception of external objects and a self-related experience (p. 246). That is, a sense of self may arise from feelings generated by control-oriented, rather than epistemic, predictive processing.

Seth's approach could be a promising avenue for a neuropsychological criterion for distinguishing the extended self from other kinds of extended cognition. The brain may interface with technology in multiple ways, including either via epistemic or control-oriented mechanisms. In the latter case, it may be reasonable to think that the extension becomes richly embodied or "lived" in a way that it doesn't in the former. Whether this requires that the processes be properly interoceptive would be a further question. If so, Seth's criterion may be too restrictive – actions using the technology might then have to be governed by specific and dedicated neural circuits that are not particularly plastic in their range of functions (e.g., brainstem regulation of heart rate), and this kind of brain-technology interface may be limited to certain medical contexts. But if the control-oriented predictive processing for the extended self can include proprioceptive mechanisms – mechanisms that predict and control the position of the body – then technological extensions might become part of the body and hence part of an extended embodied self. This idea is not without problems. Seth's framework is designed around interoception and specifically excludes proprioception on grounds that the former but not the latter is a homeostatic function that is not designed to gain knowledge of causal structure but rather to maintain the visceral system. That is, Seth categorizes proprioception with perception rather than with the lived inner sense of a "beast machine." This may well be right, and if it is then technologically extended selves will be rare indeed. However, it remains highly intuitive that proprioception is, properly speaking, *self*-perception. Certainly, this does not settle the matter. But it does provide some prima facie plausibility for the inclusion of proprioception in a modified version of Seth's framework.

Even so, this criterion for the extended self is still quite demanding. It excludes most cases of extended cognition that otherwise pass the test of the parity principle, and it remains stronger than comparable ideas in the literature, such as the proposal by Reiner and Nagel (2017), discussed in Section 2.2 above. Recall that they propose a criterion for "technologies of the extended mind" (TEM) that is intended to strengthen the bare parity principle. Roughly, they propose that a technology becomes part of a person's mind when it is no longer questioned by the person and the interaction becomes relatively seamless, such that the technology is perceived by the subject *as* an extension of their mind. This is progress, insofar as Reiner et al argue that something more than mere parity is needed. But it remains an account of extended cognition, stopping short of drawing a distinction between extended cognition and extended self. Instead, they offer a distinction between TEMs and merely *assistive* or supporting technologies (2017, p. 110). Assistive technologies, they hold, are those that may pass the test of parity but not the stronger requirement of "relatively seamless" interaction. As an example, they present the case of John the cab driver. When John first uses the GPS, the technology is merely assistive, but when he comes to trust it unthinkingly and automatically, it becomes a proper TEM. Their approach is thus notably more permissive than the present proposal. We suggest that the key difference lies, not in seamless or unquestioned interaction, but in embodiment grounded in control-oriented neural processing. We agree that John's GPS should, under some circumstances, be understood as extended cognition. But we question whether it is usefully understood as part of John, under any circumstances. It is not a case of an extended self.

But return to the Inga case with the telepresence robot, introduced in Section 2.3 above. If the entwinement of Inga's brain with the technology really does reach the third level described, (X-System: limbic-ventral striatal loop), this may plausibly entrain the neural mechanisms for proprioception, indicating that the robot has really become Inga's body – or perhaps even that Inga's brain has become part of the robot. Extended selves may also begin to crop up in the metaverse or other virtual and extended reality platforms in which virtual embodiment becomes a genuine habitus. Cases like these might illustrate the advantages of the concept of an extended self as we've introduced it here because the "entwining" relation between the person and the technology is specifically grounded in the neural roots of the embodied self, rather than in the phenomenology of body tech. This allows for remote and virtual extensions of the embodied self, rather than being limited to things like prostheses, glasses, or suits. Classical treatments, such as Merleau-Ponty's example of the blind man's cane, are indeed still going to be examples, but they are only one kind of extended self in a genre that is unified via control-oriented mechanisms for interoception and proprioception.

If sustainable, the distinction between extended cognition and extended self may prove useful for writers working on the ethical and policy dimensions of new technology. For cases of extended cognition without an extended self, it may be that existing policy and case law, developed for respecting privacy and personal data in, e.g., the software industry or medical technology, could be applied fairly directly. In those cases, providers and third parties certainly have obligations to the individual user. But users remain responsible for knowing the terms and conditions, and for realizing that they voluntarily enter a quasi-public or market-like space. Thus, extended cognition (without self-extension) might have the same general moral and legal status as other kinds of personal data – perhaps at one end of a spectrum of such data. Cases that do involve a genuinely extended self, however, might raise more interesting questions. Here, infringements by others may be understood more on the model of violence or assault. For instance, there seems to be a right to self-presentation, such that people can control the way they present themselves to others. For example, individuals can present themselves as masculine or feminine. On the other hand, people cannot control the way others perceive them or think about them. How would this right to self-presentation transfer to an extended self? Would a third-party intervention or control of the way a virtual self appears in a virtual space constitute an assault? More generally, what are the relations between a person and a self, such that the rights and interests of the former are also those of the latter? This would be a difficult question in any context but may become particularly acute in the context of an extended self, in which the self no longer clearly coincides with the person or subject.

Conclusions

As XR platforms evolve into algorithmic devices that seamlessly interface with neurocognitive mechanisms, methodological, ethical, and conceptual issues arise. From the perspective of neuropsychological assessment, XR technology may transform both experimental and clinical practice through a balanced combination of improved ecological validity and experimental control. Received wisdom derived from low-dimensional techniques may be updated, while at the same time, researchers must face new concerns about the effects of technology on the brains of users. At the same time, familiar ethical problems from biomedical and legal contexts also arise at the intersection of XR technology and neuropsychology. Algorithmic devices that extend or assist cognition may threaten privacy or undermine the agency. And if persons are more intimately linked to their technology, then removing or damaging these links may be tantamount to a violent crime.

Our discussion of neurocognitive extension in XR contexts also attempts to build on Chalmers' (2019) argument that cognition can extend, not just across the boundaries of skin and skull, but also across the bounds of perception and action. One implication of Chalmers' approach is that, in these cases, cognition can take place "outside" or "beyond" the conscious subject whose cognitions they are. We follow Chalmers here, but also attempt to open a third category, that of the self, which does not perfectly coincide with either cognition or consciousness. Embodiment – the lived body – might be capable of sensorimotor transfiguration that subverts what perception and action consist of, in the first place. Does the telepresence robot simply alter the locus of perception and action, such that the self still resides within these bounds? Dennett's (1978) characteristically deflationary discussion of this sort of case remains equivocal – pointing perhaps to eliminativism about the self or to an identification of the self with the neural correlates of consciousness. But perhaps this is too quick? Perhaps there really are embodied selves that can extend across the brain, animal body, and technology. Whatever future research may show, Dennett was surely right that the way forward lies through the brain.

References

Abbott, E. A. (1952). *Flatland: A Romance of Many Dimensions: With Illus. by the Author, a Square*. Dover Publications.
Adams, F., & Aizawa, K. (2001). The bounds of cognition. *Philosophical Psychology, 14*(1): 43–64.

Baars, B. (1988). *A Cognitive Theory of Consciousness*. Cambridge University Press.

Beauchamp, M. H. (2017). Neuropsychology's social landscape: Common ground with social neuroscience. *Neuropsychology, 31*(8): 981.

Behr, K. M., Nosper, A., Klimmt, C., & Hartmann, T. (2005). Some practical considerations of ethical issues in VR research. *Presence, 14*(6), 668–676.

Bohil, C. J., Alicea, B., & Biocca, F. A. (2011). Virtual reality in neuroscience research and therapy. *Nature Reviews Neuroscience, 12*(12): 752–762.

Bréchet, L., Mange, R., Herbelin, B., Theillaud, Q., Gauthier, B., Serino, A., & Blanke, O. (2019). First-person view of one's body in immersive virtual reality: Influence on episodic memory. *PLoS One, 14*(3): e0197763.

Burgess, P. W., Alderman, N., Forbes, C., Costello, A., LAURE, M. C., Dawson, D. R., ... & Channon, S. (2006). The case for the development and use of "ecologically valid" measures of executive function in experimental and clinical neuropsychology. *Journal of the International Neuropsychological Society, 12*(2): 194–209.

Cabeza, R., Prince, S. E., Daselaar, S. M., Greenberg, D. L., Budde, M., Dolcos, F.,... & Rubin, D. C. (2004). Brain activity during episodic retrieval of autobiographical and laboratory events: an fMRI study using a novel photo paradigm. *Journal of Cognitive Neuroscience, 16*(9): 1583–1594.

Cabeza, R., and St Jacques, P. (2007). Functional neuroimaging of autobiographical memory. *Trends in Cognitive Sciences, 11*: 219–227.

Cai, C., Yuan, K., Yin, J., Feng, D., Bi, Y., Li, Y.,... & Tian, J. (2016). Striatum morphometry is associated with cognitive control deficits and symptom severity in internet gaming disorder. *Brain Imaging and Behavior, 10*: 12–20.

Chalmers, D, (2019). Extended cognition and extended consciousness. In Columbo, Irvine, & Stapleton (eds.), *Andy Clark and His Critics*. Oxford University Press: 9–20. https://doi.org/10.1093/oso/9780190662813.001.0001

Clark, A. (2008). *Supersizing the Mind: Embodiment, Action, and Cognitive Extension*. OUP USA.

Clark, A. (2009). Letter to the editor. *London Review of Books 31.6* (March 26). https://www.lrb.co.uk/the-paper/v31/n06/letters

Clark, A. (2015). *Surfing Uncertainty: Prediction, Action, and the Embodied Mind*. Oxford University Press.

Clark, A., & Chalmers, D. (1998). The extended mind. *Analysis, 58*(1): 7–19.

Craig, A. D. (2009). How do you feel—now? The anterior insula and human awareness. *Nature Reviews Neuroscience, 10*(1): 59–70.

Dehaene, S. (2014). *Consciousness and the Brain: Deciphering How the Brain Codes Our Thoughts*. Penguin Books.

Dennett, D.C. (1978). Where Am I? In *Brainstorms: Philosophical Essays on Mind and Psychology*. Montgomery, VT: Bradford Books: 310–323. https://mitpress.mit.edu/9780262540377/brainstorms/

Dennett, D. C. (1996). *Kinds of Minds*. New York: Basic Books.

Ding, W. N., Sun, J. H., Sun, Y. W., Chen, X., Zhou, Y., Zhuang, Z. G.,... & Du, Y. S. (2014). Trait impulsivity and impaired prefrontal impulse inhibition function in adolescents with internet gaming addiction revealed by a Go/No-Go fMRI study. *Behavioral and Brain Functions, 10*(1): 1–9.

Dombeck, D. A., & Reiser, M. B. (2012). Real neuroscience in virtual worlds. *Current Opinion in Neurobiology, 22*(1): 3–10.

Fitz, N. S., & Reiner, P. B. (2016). Perspective: Time to expand the mind. *Nature, 531*(7592): S9–S9.

Friston, K. (2010). The free-energy principle: A unified brain theory? *Nature Reviews Neuroscience, 11*(2): 127–138.

Gallagher, S. (2013). A pattern theory of self. *Front Hum Neuroscience, 7*: 443.

Hayles, N. K. (2017). *Unthought: The Power of the Cognitive Nonconscious*. Chicago University Press.

Holler, D. E., Fabbri, S., & Snow, J. C. (2020). Object responses are highly malleable, rather than invariant, with changes in object appearance. *Scientific Reports, 10*(1): 4654

Hohwy, J. (2013). *The Predictive Mind*. OUP Oxford.

Hou, H., Jia, S., Hu, S., Fan, R., Sun, W., Sun, T., & Zhang, H. (2012). Reduced striatal dopamine transporters in people with internet addiction disorder. *Journal of Biomedicine and Biotechnology, 2012*: 854524. Doi:10.1155/2012/854524. https://www.hindawi.com/journals/bmri/2012/854524/

Hyde, J. S. (2005). The gender similarities hypothesis. *American Psychologist, 60*(6): 581.

James, W. (2007 [1890]). *The Principles of Psychology* (Vol. 1). Cosimo, Inc.

Jeong, B. S., Han, D. H., Kim, S. M., Lee, S. W., & Renshaw, P. F. (2016). White matter connectivity and Internet gaming disorder. *Addiction Biology, 21*(3): 732–742.

Jin, C., Zhang, T., Cai, C., Bi, Y., Li, Y., Yu, D.,... & Yuan, K. (2016). Abnormal prefrontal cortex resting state functional connectivity and severity of internet gaming disorder. *Brain Imaging and Behavior, 10*: 719–729.

Jolly, E., & Chang, L. J. (2019). The flatland fallacy: Moving beyond low–dimensional thinking. *Topics in Cognitive Science, 11*(2): 433–454.

Kane, R. L., & Parsons, T. D. (eds.) (2017). *The Role of Technology in Clinical Neuropsychology*. Oxford University Press.

Kennedy, D. P., & Adolphs, R. (2012). The social brain in psychiatric and neurological disorders. *Trends in Cognitive Sciences, 16*(11): 559–572.

Kim, S. H., Baik, S. H., Park, C. S., Kim, S. J., Choi, S. W., & Kim, S. E. (2011). Reduced striatal dopamine D2 receptors in people with Internet addiction. *Neuroreport, 22*(8): 407–411.

Kisker, J., Gruber, T., & Schöne, B. (2021a). Experiences in virtual reality entail different processes of retrieval as opposed to conventional laboratory settings: A study on human memory. *Current Psychology, 40*: 3190–3197.

Kisker, J., Gruber, T., & Schöne, B. (2021b). Virtual reality experiences promote autobiographical retrieval mechanisms: Electrophysiological correlates of laboratory and virtual experiences. *Psychological Research, 85*: 2485–2501.

Klein, E. (2017). Neuromodulation ethics: Preparing for brain-computer interface medicine. In Illes (ed.), *Neuroethics: Anticipating the Future*. Oxford University Press.

Ko, C. H., Hsieh, T. J., Chen, C. Y., Yen, C. F., Chen, C. S., Yen, J. Y.,... & Liu, G. C. (2014). Altered brain activation during response inhibition and error processing in subjects with internet gaming disorder: A functional magnetic imaging study. *European Archives of Psychiatry and Clinical Neuroscience, 264*: 661–672.

Ko, C. H., Hsieh, T. J., Wang, P. W., Lin, W. C., Yen, C. F., Chen, C. S., & Yen, J. Y. (2015). Altered gray matter density and disrupted functional connectivity of the amygdala in adults with Internet gaming disorder. *Progress in Neuro-Psychopharmacology and Biological Psychiatry*, 57: 185–192.

Ko, C. H., Liu, G. C., Hsiao, S., Yen, J. Y., Yang, M. J., Lin, W. C.,... & Chen, C. S. (2009). Brain activities associated with gaming urge of online gaming addiction. *Journal of Psychiatric Research*, 43(7): 739–747.

Kriegel, U. (2009). *Subjective Consciousness: A Self-representational Theory*. Oxford: OUP Oxford.

Kühn, S., Romanowski, A., Schilling, C., Lorenz, R., Mörsen, C., Seiferth, N.,... & Gallinat, J. (2011). The neural basis of video gaming. *Translational Psychiatry*, 1(11): e53–e53.

Levine, J. (2001). *Purple Haze: The Puzzle of Consciousness*. Oxford University Press.

Lochhead, I., Hedley, N., Çöltekin, A., & Fisher, B. (2022). The immersive mental rotations test: Evaluating spatial ability in virtual reality. *Frontiers in Virtual Reality*, 3: 5.

Lv, Z. (2020). Virtual reality in the context of internet of things. *Neural Computing and Applications*, 32(13): 9593–9602.

Monahan, J. S., Harke, M. A., & Shelley, J. R. (2008). Computerizing the mental rotations test: Are gender differences maintained? *Behaviour Research Methods*, 40(2): 422–427.

Nagel, S. K. & Reiner, P. B. (2018). Skillful use of Technologies of the Extended Mind illuminate practical paths towards an ethics of consciousness. *Frontiers in Psychology*, 9: 1–2.

Neisser, J. (2015). *The Science of Subjectivity*. Palgrave Macmillan Press.

Neisser, U. (1978). Memory: What are the important questions? In M. M. Gruneberg, P. E. Morris, & R. N. Sykes (eds.), *Practical Aspects of Memory*, pp. 3–24. San Diego: Academic Press.

Niu, X., Gao, X., Zhang, M., Yang, Z., Yu, M., Wang, W.,... & Zhang, Y. (2022). Meta-analysis of structural and functional brain alterations in internet gaming disorder. *Frontiers in Psychiatry*, 13: 1–14.

Noël, X., Brevers, D., & Bechara, A. (2013). A neurocognitive approach to understanding the neurobiology of addiction. *Current Opinion in Neurobiology*, 23(4): 632–638.

Nosek, B. A., & Errington, T. M. (2020). What is replication? *PLoS Biology*, 18(3): e3000691.

Nosek, B. A., Hardwicke, T. E., Moshontz, H., Allard, A., Corker, K. S., Dreber, A., & Vazire, S. (2022). Replicability, robustness, and reproducibility in psychological science. *Annual Review of Psychology*, 73: 719–748.

Open Science Collaboration. (2015). Estimating the reproducibility of psychological science. *Science*, 349(6251): aac4716.

Palmas, F., & Klinker, G. (2020, July). Defining extended reality training: a long-term definition for all industries. In *2020 IEEE 20th International Conference on Advanced Learning Technologies (ICALT)*: 322–324. IEEE.

Parsons, T. D. (2015). Virtual reality for enhanced ecological validity and experimental control in the clinical, affective and social neurosciences. *Frontiers in Human Neuroscience*, 9: 660.

Parsons, T. D. (2017). *Cyberpsychology and the Brain: The Interaction of Neuroscience and Affective Computing*. Cambridge University Press.

Parsons, T. D. (2019). *Ethical Challenges in Digital Psychology and Cyberpsychology*. Cambridge University Press.

Parsons, T. D. (2021). Ethical challenges of using virtual environments in the assessment and treatment of psychopathological disorders. *Journal of Clinical Medicine*, 10(3): 378.

Parsons, T. D., Carlew, A. R., Magtoto, J., & Stonecipher, K. (2017). The potential of function-led virtual environments for ecologically valid measures of executive function in experimental and clinical neuropsychology. *Neuropsychological Rehabilitation*, 27(5): 777–807.

Parsons, T. D., & Duffield, T. (2019). National Institutes of Health initiatives for advancing scientific developments in clinical neuropsychology. *The Clinical Neuropsychologist*, 33(2): 246–270.

Parsons, T., & Duffield, T. (2020). Paradigm shift toward digital neuropsychology and high-dimensional neuropsychological assessments. *Journal of Medical Internet Research*, 22(12): e23777.

Parsons, T. D., Gaggioli, A., & Riva, G. (2017). Virtual reality for research in social neuroscience. *Brain Sciences*, 7(4): 42.

Parsons, T. D., Gaggioli, A., & Riva, G. (2020). Extended reality for the clinical, affective, and social neurosciences. *Brain Sciences*, 10(12): 922.

Parsons, T., Larson, P., Kratz, K., Marcus, T., Bluestein, B., & Galen Buckwalter, J., et al. (2004). Sex differences in mental rotation and spatial rotation in a virtual environment. *Neuropsychologia*, 42(4): 555–562.

Pashler, H., & Wagenmakers, E. (2012). Editors' Introduction to the Special Section on Replicability in Psychological Science: A Crisis of Confidence? *Perspectives on Psychological Science*, 7(6): 528–530.

Price, C. J. (2018). The evolution of cognitive models: From neuropsychology to neuroimaging and back. *Cortex*, 107: 37–49.

Ramirez, E. J. (2019). Ecological and ethical issues in virtual reality research: A call for increased scrutiny. *Philosophical Psychology*, 32(2): 211–233.

Ramirez, E. J., & LaBarge, S. (2018). Real moral problems in the use of virtual reality. *Ethics and Information Technology*, 20: 249–263.

Ramirez, E. J., & LaBarge, S. (2020). Ethical issues with simulating the bridge problem in VR. *Science and Engineering Ethics*, 26: 3313–3331.

Reiner, P. B. & Nagel, S. K. (2017). Technologies of the extended mind: Defining the issues. In J. Illes & S. Hossain (eds.), *Neuroethics: Anticipating the Future*, pp. 108–122. Oxford: Oxford University Press.

Sauzéon, H., Pala, P. A., Larrue, F., Wallet, G., Déjos, M., Zheng, X., ... & N'Kaoua, B. (2012). The use of virtual reality for episodic memory assessment. *Experimental Psychology*, 59(2): 99–108

Schiebener, J., & Brand, M. (2015). Decision making under objective risk conditions—a review of cognitive and emotional correlates, strategies, feedback processing, and external influences. *Neuropsychology Review*, 25: 171–198.

Schöne, B., Wessels, M., & Gruber, T. (2019). Experiences in virtual reality: A window to autobiographical memory. *Current Psychology, 38*(3): 715–719.

Seth, Anil K. (2019). Being a beast machine: The origins of selfhood in control-oriented interoceptive inference. In Columbo, Irvine & Stapleton (eds.), *Andy Clark and His Critics*. Oxford University Press: 238–253. https://doi.org/10.1093/oso/9780190662813.001.0001

Slater, M., Gonzalez-Liencres, C., Haggard, P., Vinkers, C., Gregory-Clarke, R., Jelley, S.,... & Silver, J. (2020). The ethics of realism in virtual and augmented reality. *Frontiers in Virtual Reality, 1*: 1.

Solly, J. E., Hook, R. W., Grant, J. E., Cortese, S., & Chamberlain, S. R. (2022). Structural gray matter differences in problematic usage of the internet: A systematic review and meta-analysis. *Molecular Psychiatry, 27*(2): 1000–1009.

Stanovich, K. E. (2009a). Distinguishing the reflective, algorithmic, and autonomous minds: Is it time for a tri-process theory. In J. Evans & K. Frankish (eds.), *Two Minds: Dual Processes and Beyond*, pp. 55–88. Oxford: Oxford University Press.

Sun, J. T., Hu, B., Chen, T. Q., Chen, Z. H., Shang, Y. X., Li, Y. T.,... & Wang, W. (2023). Internet addiction-induced brain structure and function alterations: A systematic review and meta-analysis of voxel-based morphometry and resting-state functional connectivity studies. *Brain Imaging and Behavior, 17*(3): 329–342

Turel, O., Serenko, A., & Giles, P. (2011). Integrating technology addiction and use: An empirical investigation of online auction users. *MIS Quarterly, 35*(4): 1043–1061.

Tye, M. (1995). *Ten Problems of Consciousness: A Representational Theory of the Phenomenal Mind*. MIT Press.

Vass, L. K., Copara, M. S., Seyal, M., Shahlaie, K., Farias, S. T., Shen, P. Y., & Ekstrom, A. D. (2016). Oscillations go the distance: low-frequency human hippocampal oscillations code spatial distance in the absence of sensory cues during teleportation. *Neuron, 89*(6): 1180–1186.

Voyer, D., Voyer, S., & Bryden, M. P. (1995). Magnitude of sex differences in spatial abilities: A meta-analysis and consideration of critical variables. *Psychological bulletin, 117*(2): 250–270.

Wilms, M., Schilbach, L., Pfeiffer, U., Bente, G., Fink, G. R., & Vogeley, K. (2010). It's in your eyes—using gaze-contingent stimuli to create truly interactive paradigms for social cognitive and affective neuroscience. *Social Cognitive and Affective Neuroscience, 5*(1): 98–107.

Wood, S. M. W., & Bechara, A. (2014), The neuroscience of dual (and triple) systems in decision making. In V. F. Reyna & V. Zayas (eds.), *The Neuroscience of Risky Decision Making*, pp. 177–202. Washington, DC: American Psychological Association.

Wu, J. W., Chou, D. W., & Jiang, J. R. (2014, September). The virtual environment of things (veot): A framework for integrating smart things into networked virtual environments. In *2014 IEEE International Conference on Internet of Things (iThings), and IEEE Green Computing and Communications (Green-Com) and IEEE Cyber, Physical and Social Computing (CPSCom)*: 456–459. IEEE.

Yuan, K., Qin, W., Yu, D., Bi, Y., Xing, L., Jin, C., & Tian, J. (2016). Core brain networks interactions and cognitive control in internet gaming disorder individuals in late adolescence/early adulthood. *Brain Structure and Function, 221*: 1427–1442.

Zaki, J., & Ochsner, K. (2009). The need for a cognitive neuroscience of naturalistic social cognition. *Annals of the New York Academy of Sciences, 1167*(1): 16–30.

Zhang, Y., Mei, W., Zhang, J. X., Wu, Q., & Zhang, W. (2016). Decreased functional connectivity of insula-based networks in young adults with internet gaming disorder. *Experimental Brain Research, 234*: 2553–2560.

Zhao, H., Turel, O., Bechara, A., & He, Q. (2023). How distinct functional insular subdivisions mediate interacting neurocognitive systems. *Cerebral Cortex, 33*(5): 1739–1751.

Part 2

Is There an Ethics for Extended Reality?

5   Mediated Reality

Michael Madary

Introduction: All Intentionality Is Mediated

The existence of this volume is evidence that extended reality media technology is deemed worthy of philosophical attention. It is worthy of this attention because it is far more powerful than more familiar media forms. But not only that. The view that I will be advancing in this chapter is that *all* media technology is worthy of additional philosophical attention. In order to appreciate the metaphysical, psychological, and ethical significance of extended reality (XR) media technology, it helps first to appreciate the significance of media technology in general. Thus, the goal of this chapter is to move toward a better understanding of XR technology by broadening the scope of inquiry far beyond XR technology itself. My strategy will be to focus on media, in its broadest sense, and its relations with human thought in the broadest sense.

Let us begin with the philosophical concept of *intentionality*. It is the technical philosophical term for all human thought, in the broad sense to include states such as perception and memory. The term does not refer to having an intention or not, as in, did you intend to hide the Oculus from little Irma or did you merely misplace it? Rather, the term intentionality refers to the property of all mental states whereby they are about something or directed at something. My belief that there is milk in my refrigerator is an intentional state directed towards the milk that is in my refrigerator. Intentionality is a powerful philosophical concept because of its diverse applications to all faculties of mind. There are many ways of approaching and analyzing intentionality in the philosophical literature. The way that I choose here is a heuristic approach that is meant to offer a technique for analyzing intentional states with neither an overly demanding conceptual apparatus nor any sort of metaphysical commitment.[1] In particular, I delineate three general regions of intentional objects – the three broad regions about which we can have thoughts. Those domains are (1) the world; (2) oneself; (3) others. Alongside these three domains, I will also include

DOI: 10.4324/9781003359494-8

the valence or values that tend to accompany intentional states. That is, we have positive or negative feelings about elements of the world, oneself, or other people.

This approach is motivated by the conviction that humanity in general, and philosophers in particular, have drastically underestimated the ways that different media technologies shape and determine intentional states. The Western philosophical tradition has been concerned with our endeavor to make reality more intelligible, but it has largely omitted from its analysis the tools that we use to do so. Media technologies are tools of intelligibility.

Thus my strategy is to draw out the distinctly powerful features of XR media by first presenting reflections upon the relationship between intentionality and media. My main claim about that relationship is that *all intentionality is mediated*. The reason why all intentionality is mediated is that the living body itself is the primary medium. All media technology expands the possibilities of intentionality beyond the mediation of the living perceiving body. In the following section of this chapter, I will make the case, following Aristotle and Husserl, that the living body is the primary medium.

After making the case that the body is the primary medium (Section 2), I will turn to media technology in Sections 3 and 4. Section 3 will cover the most important media technology in the history of thought: the written word. Section 4 will focus on the topic of this volume itself, XR technologies such as virtual reality (VR) and augmented reality (AR).

Then in Section 5, this chapter will take an applied turn. By that point, I will have made the case that there is something special about XR, but also that there is something special about all media technologies. Media shape the mind and XR has the potential to do so in an especially overpowering way. In Section 5, I will address the foreseeable risks of XR technology and make the case that we all have the right to avoid the encroachment of XR into our lives (and onto our bodies). There I will draw an analogy to the present situation with the case of the personal automobile. Just as the industry created pressure for the widespread adoption of the automobile, the immersive technology industry is poised to do the same for their product today.

Organic Reality: The Body Is the Primary Medium

The ease with which we perceive the world outside of ourselves can lead one to think about perception as a one-way causal direction. We open our eyes and we see a world. This one-directional flaw is manifest in the common claim that our perceptual states are caused by the outside world. For example, in his development of a realist position with regard to virtual objects, David Chalmers relies on this flawed assumption in his premise that "The objects that we perceive are the causal basis of our perceptual experiences" (2017: 318). There is a sense, of course, in which our perceptional states are indeed caused by the outside world. But that is an incomplete and philosophically misleading picture of the perceptual relationship. For the perceptual states that are caused by the outside world are always the only states that can be enabled by our living perceiving bodies. Bees can see light in the ultraviolet spectrum while humans cannot. This difference is one that is based on bodily details – we do not have the same sorts of photoreceptors that bees do. It would be more accurate to say that perceptual states are the result of the interaction between the organism's environment and the organism's bodily possibilities for sensations. The body undergoes changes in response to the environment that enable perceptual states to arise. The details of one's body determine the possible range of perceptual states that one can have. In addition to the details of the sense organs of one's body, there are other factors that determine perceptual processing, such as the contextual preferences of the organism, and relatedly, perceptual anticipation (or predictive signals). In these ways, the living body mediates the perceptual states of organisms.[2]

According to Friedrich Kittler, the etymological origin of the term 'media' can be found at the origin of the Western philosophy of perception. In his treatment of perception in *De Anima*, Aristotle took the Greek preposition for 'between' (*metaxú*) and created the noun (*tò metaxú*) from which we have 'media' in English today (Kittler 2009: 26). In contrast to the ancient atomists who considered perceptual signals to travel through the void to reach our eyes and ears, Aristotle insisted that all perceptual modalities require a medium, a *between* through which the perceptual relation obtains (Kittler 2009: 25–26). For Aristotle, the primary sense modality is touch and the medium of touch, according to him, is the living body itself (*De Anima* 422b20). Bringing in promising work from the 21st century, we might broaden Aristotle's concept of touch to include awareness of the body itself through what is now known as interoception (Seth 2013).

In a further development of this theme, Edmund Husserl claims that "The body is, in the first place, the *medium of all perception*; it is the *organ of perception* and is *necessarily* involved in all perception" (Husserl 1989: §18b). To illustrate this claim, he offers the example of a finger that is injured through a burn. We experience the abnormality of tactile sensation using the burned finger precisely because the medium is altered in a noticeable way. Husserl also used the example of the ingestion of the drug santonin as an illustration of how changing the living body, changing the medium of perception, changes our pattern of visual experience. Santonin was taken in the early 20th century to treat intestinal parasites. A side-effect of santonin is the experience of a yellow tint to one's entire field of vision. The drug changes the visual medium which results in an alteration to the visual experience. In both the example of the burned finger and the ingestion of santonin, we experience perceptual abnormality as a departure from normal bodily perception. Normal perception is, Husserl argues, constituted by an intersubjective community of perceivers (Husserl 1973: vol. XIV: 133n; Wehrle 2015; Madary 2019). Importantly, it is the similarity with the medium of perception, the living body, that enables perceptual normality to be shared within the community. Perceptual norms are always relative to communities of living embodied organisms. This relativity entails that norms constitutively depend upon sufficient similarity among bodily perceptual abilities. If there were sufficient diversity among bodily abilities, then there would be no norms. Some cases of bodily diversity, such as loss of sight, can place limitations on the sorts of perceptual experiences that can be shared intersubjectively. The important point, though, is that the living body is the primary medium of perception for all of us – regardless of the degree to which our living body shares in or diverges from perceptual normality.

Since XR such as immersive virtual reality depends so heavily on the relationship between action and visual perception, it will be especially appropriate here to offer a final example of the body as the medium by considering the cycle of action and visual perception. As I have argued elsewhere (2017), visual experience involves an ongoing process of anticipation and fulfillment (this claim is first expressed in Husserl's work). This cycle typically, almost always, involves self-generated movements – actions. We move our eyes, our heads, and our body, and all of these movements offer us new perspectives on the visual scene. Vision involves the implicit anticipation of the consequences of these movements. Importantly, this cycle reveals how vision and action are interdependent in a cyclical manner. Since the sorts of actions that we can generate are dependent upon the sorts of bodies we have, the details of our embodiment shape the possibilities of visual experience.

This interdependence of action and perception is especially important for our purposes here because the media of XR using a head-mounted display works by tracking the cycle of action and perception in the primary medium of the body. The head-mounted display tracks our movements and then updates the visual display accordingly. When we look down, the visual display shows us the virtual world beneath our virtual selves. When we look up, we see the virtual world above our heads. As we will explore in Section 4, XR mediation is so powerful and philosophically interesting because it *lays an electronic medium directly over the primary metabolic medium.*

One of our primary modes for distinguishing the objective (the world) from the subjective (the self) is through the interaction of movement and vision. When visual experiences do not change in the way that we expect as a result of our self-generated movements then we have reason to suspect that our experience is not of the outside world. A good example of this occurrence is the case of phosphenes (commonly known as "seeing stars"). When you stand up too fast and experience phosphenes darting around peripheral vision, you have implicit awareness that something is odd – the appearance of the phosphenes is not contingent upon your self-generated actions. That is, you can turn your head and they are still there. In this way, we are able to distinguish sensations that are generated by the visual organ itself, on the one hand, and sensations that reveal the world outside us on the other hand. Susanna Siegel (2010: 179) explores this distinction using the concept of *perspectival connectedness*. According to Siegel, perspectival connectedness obtains when a substantial change of perspective on an object results in a change in visual experience. Phosphenes lack perspectival connectedness. In contrast, normal objects in the world show up for us visually as being perspectival connected because their appearances change systematically with our movements.

The technical term for the ways that appearances should change as we move is *sensorimotor contingency* (O'Regan and Noë 2001). The visual input to the eyes is contingent upon motor movements. As we develop from birth, we learn these contingencies. If we wear goggles that distort these contingencies, then we have to relearn the contingencies through practice (Degenaar 2014). Virtual reality creates the illusion of being somewhere else precisely by recreating sensorimotor contingencies through body tracking and the head-mounted display (Slater 2009).

The changes in the body or that the body undergoes in order to perceive the world are changes that can be characterized as the changes in the medium of perception, the lens through which all reality first and foundationally shows up for us. What we traditionally call media – print media, radio, television, and connected devices – are all inorganic media that open up new intentional possibilities for the organic body.[3] The possible intentional states that we have with the naked, so to speak, body are greatly expanded through inorganic mediation. For example, consider a fan of Thoroughbred racing interested in the Kentucky Derby who lives many miles away from Kentucky. Perception with the primary medium of the living body (without the help of inorganic media) will not give this horseracing fan intentional access to any details of the race. Through inorganic media, the racing enthusiast's intentional possibilities are greatly expanded. The fan can read about the race in print media, listen to prognostication about the race on the radio, watch the race live on television, and wager on the race using the internet.

Before going further into the ways that inorganic media change intentionality, let us return to the living body in order to consider how bodily perception first opens up our access to the three regions of intentionality.

Above, I mentioned how perspectival connectedness can help us to test reality, to distinguish between what is outside of the body and what is generated by the body. This fact brings us back to the three regions of intentionality introduced above. How do the cycle of action and visual perception relate back to the three regions?

The way in which the cycle of action and visual perception gives rise to experience is a way that exploits the distinction between *self* and *world*. Or rather, it is a direct and efficient way of distinguishing between self and world. When sensorimotor contingencies change in a way that shows perspectival connectedness, we take what is revealed to us to be independent of ourselves. We take it to be a revealing of the world. When we have visual experiences that do not exhibit perspectival connectedness such as phosphenes – then we take what we experience visually to be a product of the embodied self, of the visual organ. Self-generated action itself is crucial for the distinction between self and world, for we are always using the sense of agency to track actions that are self-generated from actions that are movements driven by some force from the external world. Disruptions of our ability to feel a sense of agency for our self-generated actions occur in cases of mental illness such as schizophrenia and depersonalization disorder (Frith 2005).

What about the intentional region of *other* people? There are at least two main ways in which intentional directness to the other enters into the visual experience. The first is through what is known as embodied intersubjectivity. As proponents of interactionist approaches to social cognition have long emphasized (Gallagher 2006), we first and primarily engage with others through the perception of bodily expressions, such as facial expression, gesture, and gait. The second way in which others are important for visual intentionality is through their role in objectivity itself. As many philosophers have concluded, the experience of the world as objective, as the world, is always also an experience of it as being perceivable for others. Objectivity is unthinkable without intersubjectivity.[4]

When introducing the three regions of intentionality above, I suggested that valence cuts across all of them. Valence is clearly a part of visual exploration through sensorimotor contingency, for our actions reflect the valence of what we are experiencing (Steinbock 1995: 138–139; Kelly 2005). For example, we lean in for a better look at something that is visually pleasing, or we move our gaze away from something that is discordant with our values. We reflect valence and valuation in the pattern of self-generated movements. Our movements reflect how we want to perceive the world (or what in the world we wish to perceive), and how we would like the world to be through changes that we make to the world in action.

I have made the case that the body is the primary medium. All intentionality, in all three regions, is mediated. Inorganic mediation then makes things interesting by enabling intentional states to occur that the naked body alone cannot, as we will see with one of the most important media for intentional states: the written word. We turn to this medium in the following section.

Literate Reality

The first and most famous philosophical consideration of the written word was, of course, in Plato's *Phaedrus* (275a–b). There the wise Egyptian king Thamus rejects the gift of the written word from the god Thoth. He does so for fear that this gift will cause forgetfulness in his subjects. Thamus was correct. It does cause forgetfulness by alleviating the need to remember using one's biological abilities. In the terminology being developed here, Thamus could foresee that the written word would forever change the intentional landscape for literate humans.

Apart from this famous passage, the importance of the written word for human intentional states has been largely ignored in the history of Western philosophy. According to the line of thought being developed in this chapter, this omission is a serious one. There may be important exceptions, but, as far as I can tell, the written word itself does not even begin to receive any consideration as an important philosophical topic until the 20th century. We can identify at least four distinct academic traditions in which the written word has shown up as a central theme. These traditions are: phenomenological philosophy, analytic philosophy of science, media studies, and extended cognition. In each of these traditions, we can find insightful claims about the ways in which writing technology changes human intentionality.

There is a great deal to be considered in the way that each of these academic traditions treats the written word and in the ways that these traditions might be placed into a fruitful exchange with one another. I must leave such considerations for further research. For now, here is a taste:

1 In the phenomenological tradition, Husserl came to the realization that the practice of mathematicians must depend constitutively on written symbols (1954/1970: 366). Our paradigm *a priori* discipline turns out to depend upon technology. Jaques Derrida and Bernard Stiegler develop this theme further, with Derrida (1974) giving metaphysical weight to the opposition between the written word and the spoken word and Stiegler (1998) exploring the idea that writing changes the structure of time consciousness.

2 Karl Popper, the eminent philosopher of science, distinguishes between three "worlds" (Popper 1979). The first world is that of physical objects, the second world is that of subjective mental states, and the third world

is that of objective thought. Objective thought, for Popper, depends upon the physical artifacts on which written symbols reside. Today, objective thought would involve both paper books as well as digital media. Popper, like Stiegler, was interested in the way that the third world influences the second world – how the written word changes human subjectivity. Indeed, he argued that epistemology itself should focus on the third world more than the second.[5]

3 Some of the boldest claims of a shift in human consciousness between oral and literate cultures have come from pioneering figures in media studies such as Walter Ong and Ivan Illich. By considering the history of the written word, they make the case that literacy has changed the ways that we make sense of experience itself (Ong 2012: Chapter 3) by altering, for example, the sense of self, space, and time (Illich and S-anders 1988: 41).

4 Much of the activity in cognitive science over the past couple of decades has been focused on the topics of extended or scaffolded cognition. The main idea is that human beings very often achieve our cognitive tasks through clever use of our environment (Hutchins 1995; Kirsh 1995; Sterelny 2010). By far the most famous philosophical example of extended cognition is the thought experiment of Otto, who is losing his long-term memory due to Alzheimer's disease (Clark and Chalmers 1998). He uses a notebook to record facts that he would like to remember and the argument goes that, under particular conditions, his notebook is genuinely a part of his mind. Regardless of where one stands on the boundaries of the mind, it is an important and neglected fact that Otto's use of the notebook depends upon the technology of the written word.[6] In a telling reversal of the concern that we saw in Plato's *Phaedrus*, here we have the written word taken as an enhancer, rather than destroyer, of human cognitive ability.

In all four of these academic traditions, we can find expression of the idea that the written word can have a profound impact on human thought and consciousness. Here is a brief account of some of the claims about this impact. What follows is organized according to the three regions of intentionality introduced above: world, self, and other.

With writing, *the world* itself may become the storehouse for what has traditionally been relegated to the realm of the mental. For Clark (and perhaps Stiegler), the written word is genuinely a form of memory while for Popper it is knowledge itself. Ong, citing the fieldwork of A. R. Luria (1976) among illiterate Uzbek peasants, suggests that both abstract thinking and formal logical reasoning are products of literacy. Abstract thinking enables us to organize the world according to the scientific taxonomy of genus and species. Without the permanence and organization of writing, the world may not be cognizable as having such a structure.

Similarly, if the mind is constituted by inorganic objects, then *the self* may be distributed among these objects, as Clark and Chalmers suggest in their original article proposing the extended mind hypothesis (1998: 18) and Ramirez et al. explore in this volume. Going even further, Illich and Sanders (1988: Chapter 5) have explored the ways in which the very concept of selfhood is a product of writing technology. They read various well-known autobiographies as acts of self creation.

Finally, *others* begin to have an identity that can be recorded and stored using writing technology. The ability to write is what ends the age of myth and replaces it with the possibility of history. Writing enables the state to record and document facts about each member of the collective "we" that make it up. Apart from history and statehood, writing opens up a new way of taking part in objectivity that is shared by a community of others. In particular, the permanence of writing opens up the possibility of engaging with the intentionality of others who are temporally and spatially remote from us, including those who are long gone and yet to come. The objective world that we seek to uncover through natural science is made available to us in part through the written record of what others have found. Similarly, we have the permanent possibility of leaving our own written record for posterity. Writing enables the intersubjective project of seeking objective truth to expand across generations.

These lines of thinking suggest that writing has changed permanently the way that reality shows up for us. By shaping the appearance of reality itself, writing technology also has the power to shape the values that we hold. The most obvious example of this ability is found in texts that are regarded to be sacred in religious traditions. More mundane examples would be the values engendered by the fictional stories that are shared and transmitted widely through print media.

The metaphysical, psychological, and ethical issues surrounding *any* media technology ought to be approached with the awareness of the fact that media technology is in some sense prior to metaphysical thought, prior to ethical deliberation, and prior to psychological investigation. Writing is the technological foundation for metaphysics, for all of natural science (including psychology), and for normative traditions (famously in the Decalogue).

We are right now witnessing the great transition from literate consciousness to electronic consciousness. The task that lies before us, then, is quickly to gain a better understanding of how *both* print and electronic media are intertwined with the ways that we think about ourselves, others, and the world around us. There is a great deal to consider with the rise of electronic mass media in the 20th century, but that investigation must be left for another time. The focus of this volume is the most recently developed iteration of electronic mass media and it is to this family of media that we now turn.

Electrified Reality

In the second section of this chapter, I made the case that the body is the primary medium. It accesses the visual world through the cycle of action and perception. This cycle is the basis of intentional access to reality for all animals including us, the rational animal. The third section makes the case that a large part of the rationality enjoyed by the rational animal is made possible by the technology of the written word. Electronic wearable media are different from the written word in important ways. The written word is constrained in that it is *disembodied* and it is restricted to a *narrative* structure. That is, the written word does not directly target the bodily process by which we first perceive reality, nor can it really go beyond the narrative structure – the non-linear narrative is still narrative. XR technology, in contrast, is *embodied* by directly working through the cycle of action and perception and it can have a non-narrative structure of pure *data*.

As Carr (2020) has demonstrated, the literate animal capable of "deep reading" is facing extinction. It is being replaced now by an animal with electronic intentionality. Electronic mass media has developed rapidly in the past century or so. Now it is becoming seamlessly integrated into the cycle of action and perception with wearable electronic devices of the sort that make up XR technology. (The integration of technology into the cycle has been already well underway for over a decade now with the portable smartphone.) Let us consider how wearable XR technology, specifically VR, directly engages the landscape of our three regions of intentionality: world, self, and other.

Corresponding to the region of *the world*, the place illusion (Slater 2009) is the feeling of being somewhere else, of being in a virtual world. It can be induced by enabling subjects to enjoy rich patterns of sensorimotor contingencies with the head-mounted display. Corresponding to the intentional region of *self*, the illusion of embodiment (Slater et al. 2009, Madary and Metzinger 2016) is the feeling of owning and controlling a body that is virtual. Corresponding to the region of *other*, the plausibility illusion (Slater 2009) is an enhanced feeling of presence in the virtual world due to realistic social interactions with avatars in the virtual world. The virtual can be experienced as if real by targeting the three fundamental regions of intentionality.

By targeting these three fundamental regions, XR technology opens up new ways for us to conceptualize and experience reality itself, as the written word has done for over two dozen centuries. In contrast to the well-worn metaphysical dichotomy of universal and particular that we know from the written word, XR technology opens the possibility that all the particulars in the *world* itself, including human actions, offer an infinite source of data for recording, storing, and analyzing. In the age of XR, the singular identity of one's *self* is turned into a multiplicity. There is now the possibility of trying on new virtual "selves" through social media representations (Turkle 2017) and, soon, through the embodiment of avatars in social immersive VR. The relationship between self and *others* has undergone an alteration through the permanent possibility of instantaneous connection through the medium of the internet. Since our primary mode of social engagement occurs through the living body (see Section 2), the details of the social medium are important as we transition from the early internet to XR. The last two decades of internet technology have offered constant and overwhelming social connectivity that is relatively disembodied. XR holds the promise of a sort of re-embodiment during online social engagement – but of course, the virtual body need not have the appearance of our biological body.

Prior to wearable electronic technology, our patterns of action and perception gave us access to reality in its most fundamental manifestation. Now, the cycle does two jobs that overlap. First, it gives us access to the "real life" naked version of reality, but then it also gives us access to the virtual world: the social media world, the worlds of the Metaverse, and all online content. But since we make sense of "real life" reality with the help of inorganic media, the distinction between these two domains is a superficial one. Once the access to the virtual becomes seamlessly integrated into the bodily cycle of action and perception, the virtual and the "real life" worlds become seamlessly integrated as well. Those who design wearables and create the apps for them will have the ability to modify the cycle of action and perception for everyone who uses those wearables regularly. Since the modification directly targets the cycle of action and perception, they will thus be changing the way the users access and conceptualize foundational "real life" reality as well. If this line of thought is correct, then it is crucial now to be vigilant against the abuse of this power. This concern brings us to the final section of this chapter.

Oppressive Reality

So far I have made the case for three main claims. The *first* claim is that all intentionality is mediated. The body is the primary medium through which we access reality in its most direct and immediate form. The *second* claim is that our technology enables us to open up forms of intentionality that go beyond what we can access with our un-augmented bodies. These media technologies have a feedback effect on what we experience through the cycle of action and perception. By changing the way we conceptualize reality, they change what we encounter when we perceive reality. We can thus think of the intentional possibilities for any particular historically situated human being as always an entire package of living body plus media technologies. Becoming literate makes changes that depend in detail upon the level of literacy and the sorts of literature that one frequently "consumes." *Third*, I have shown that XR technology is especially powerful because it modulates directly the primary medium. Print media and even non-wearable electronic media such as television are relatively disembodied. Once we start wearing our electronic media and have them incorporated into the cycle, then we have alterations to the primary medium itself. This powerful media technology can create new fundamental changes to intentional states in all three of the regions of intentionality.

Change itself is not alarming, nor is it a powerful medium. But what *is* concerning is the pressure to adopt this new medium by forces that are poised to use it in ways that are hostile or oppressive to flourishing. Here, I make the case that this situation is the present one. In both the primary medium of the body and the media of writing technology, I have suggested that values accompany our experience in the three domains of intentionality. What are the values of the corporations, such as Facebook/ Meta, designing and maintaining XR technologies? Do they promote human flourishing or are they hostile and oppressive? These questions will be especially important to answer if there is pressure for widespread adoption of XR since widespread adoption offers opportunities for abuse. It is not clear at this point whether and how XR might be put to use by the masses. In this final section of the chapter, I present a possible undesirable course of events in order to warn against it.

Let us begin with a recent history of XR. Commercial VR was released around 2016. There was great hype about the release of these products along with market predictions that they will be eagerly purchased by the general public. These predictions turned out to be wrong. The market for VR was and remains a niche market of serious gamers. Soon after the commercial release of the products, consultants and market gurus were asking about the "killer app" for HMDs. How can we entice the grandmas of the general public to purchase the HMD as we succeeded in having grandma feel the need to purchase the smartphone? A big problem was that people simply did not seem to want it.[7]

Now, if things remain this way – and the adoption of immersive technology is limited to those (relatively few) who freely choose to use it regularly – then we can avoid the big concern that I wish to highlight. The big concern is that there will be usage and adoption of HMDs and similar XR technology due to active pressure from the industry. This has happened before in the history of technology and it looks like it may be happening now.

If we look about 100 years back in history, we can see how the industry can impose and force adoption of new and unpopular technology. My example is the personal automobile. As the automobile limits the freedom of the pedestrian body to move around the human habitat, immersive XR technology has the potential to limit the freedom of our mind through the deliberate sculpting of the intentional landscape for the interests of those creating and maintaining the virtual spaces.

As Jeff Sparrow (2019) documents, the automobile was initially not a popular technology. Very much like immersive technology, its early adoption was limited to individuals who were hobby enthusiasts with a good bit of disposable income. Less affluent urban dwellers were hostile to the new technology that was turning their streets from public spaces into danger zones. There were vigilante attacks on wealthy motorists. But the automobile industry, as we know, won the day. Through aggressive campaigning and lobbying, laws were passed that were friendly to cars and unfriendly to other modes of transportation, such as the very popular (and safer and cleaner) streetcar system in the United States. While many of us today unquestionably accept the criminalization of walking so as not to inconvenience the motorist, Sparrow points out that the automobile industry worked hard to make it so:

> To overcome the public outrage about pedestrian deaths, the industry created the idea of the "jaywalker." In the Midwest slang of the time, a "jay" was a bumpkin or a hick, a hayseed unaware of city etiquette... In the 1920s, dealers and auto clubs began using "jaywalker" for pedestrians who still believed in the old right to share the road. Local car firms paid boy scouts to distribute cards explaining the concept of jaywalking to people on the street, while the American Automobile Association promoted "safety patrols" to warn children off the street. In many places, the industry staged elaborate pageants to ridicule "jaywalkers."

What we can see with the case of the automobile is a form of technological encroachment into the lifeworld of human beings. The results have been disastrous judging by the direct death count alone – without even considering the other impact areas such as environment, personal health, and social relations.

It might turn out that there is not a similar pattern of industry driving adoption with immersive technology – indeed I hope that is the case. But there is a clear and foreseeable way in which technological encroachment may occur with immersive XR technology. It is no secret that there has been a huge amount of money invested in immersive technology, and powerful people tend to do what they can to avoid losing the money that they have invested.

One obvious route for technological encroachment of immersive XR technology is through employment and another is through education.

Employers are being told that they can improve productivity, training, and so on, by adopting immersive technology. The software company Unity commissioned a marketing report on the adoption of immersive technology during the pandemic titled "The New Way of Working is Immersive." The study states:

> While the pandemic has been a significant driver of immersive technology use, other forces will further catalyze its adoption. Decision-makers believe immersive technology can help their organizations thrive as the future of work transforms. Most of all, they predict the following of the next three years: 1) immersive technology will become a significant competitive differentiator; 2) demand for metaverse-like digital experiences will grow; and 3) spikes in systemic risks to their operations will become more common. Most agree that immersive technology can help organizations solve for areas of risk and significantly enable opportunities related to these and other future-of-work scenarios.
>
> (Forrester 2021)

These predictions are given weight because "decision-makers" make them. If these (somewhat self-fulfilling) prophesies are fulfilled, then immersive XR technology will be *normalized* through massive numbers of people using it – being pressured into using it – for the sake of keeping their jobs. In some cases, of course, employees may prefer it.

But what is the downside? In addition to the obvious physical risks of eye damage and motion sickness, the downsides are numerous (Madary and Metzinger 2016). The main downside is that XR gives great power to those who create and maintain it. This power is magnified especially by the private usage data that will be easily obtainable. Due to the use of facial and body tracking technology, the personal data that one might obtain through immersive XR technology far outstrips that which we can obtain through traditional input and output methods.

It should be noted with emphasis, of course, that the biggest corporation pushing for the adoption of immersive technology is Facebook/Meta, a corporation whose business model is to collect personal data for the purpose of large-scale behavior modification (Zuboff 2020). It should be obvious that this business model has nothing at all to do with the promotion of human flourishing. If we couple this business model with the possibility of illusions of agency when using the technology (Madary 2022), then mass adoption leans towards dystopia. Behavior modification on a large scale coupled with illusions of agency would produce a scenario in which billions of people are acting to serve the goals of the corporation without the awareness that they are doing so. The values of dignity and autonomy would no longer exist in this scenario.

As a way of illustration, recall the main points that I have been developing. Inorganic media sculpt our experience of the fundamental domains of reality, of world, self and other. With XR technology, the primary medium of the living body is distorted because the technology is worn directly upon the body. Consider the power that this can give to the corporation maintaining the XR. *Such a corporation can determine how users experience the world, how they conceive of themselves, and how they regard others.* The forms of these new determinations are difficult to predict because the medium is novel. As I have argued above, the written word was the technology that largely determined these domains for us over the past centuries. The forms of these new XR determinations will occur using a medium, unlike the written word in important ways, such as its capability to create a personalized narrative without the symbolic mediation of language. By observing current trends, here are some scenarios to consider. A corporation might like to create the desire for users to spend as much time as possible in their metaverse – perhaps to maximize spending on purchases of virtual (or physical) objects that will in no way improve quality of life. A corporation might erode and distort direct human connection by convincing users that social interactions are superior through the lens of XR – even down to the most intimate moments of our lives. A corporation might distract the populace from positive sociopolitical change with the use of a steady feed of base pleasures combined with the reinforcement of partisan polarization. Trends in these directions are present with existing media technology. I see no reason to think that XR will be any better and many reasons to think that it could become much worse.

We all have the right to refuse participation in the metaverse, to resist the emerging pressures from various industries, especially when the collection of personal data is still being practiced. A practical first step would be the regulation of XR technology in the name of occupational safety. A more ambitious (and important) step will require an attitude adjustment regarding economic growth through technological innovation. That is, technological innovation is not desirable in itself. Innovation should be always primarily in the service of promoting the dignity of the human person.

Conclusion

I would like to close by summarizing the connection between the more theoretical early sections of this chapter and the applied theme of the previous section. The main theoretical claim of this chapter is that all intentionality – all human conscious thought – is mediated. The mediation is organic in its bodily origin but then becomes partially inorganic with the use of print and electronic media technology. It is a mistake, I suggest, to assume that these various forms of mediation are entirely distinct from one another. They are not entirely distinct because our awareness of reality as it shows up in the regions of the world, self, and others is a combined result of all the mediation that we encounter through our individual historical context – both organic and inorganic mediation. The historical context now emerging is one in which a powerful electronic medium is being worn to cover the primary organic medium of the living body. There exist forces pressing for the adoption of this medium that are motivated by avarice at the expense of human flourishing. We must recognize and resist these forces. The way that *reality itself* appears and is intelligible to us and our descendants is at stake.

Notes

1 The modern conception of intentionality famously originates in Brentano (2015). A classic treatment of intentionality for the tradition of phenomenological philosophy can be found in Husserl's fifth and sixth Logical Investigations (Husserl 1900/2001, also see Madary 2012). Much of the debate around intentionality has to do with various attempts to explain intentionality in a naturalistic manner (such as in Fodor 1987 and Millikan 1987). I take no position here with regard to the naturalization project.
2 The following works address some of the ways that the context of the perceiver can influence perceptual processing: Freeman (1960), Merleau-Ponty (1964), Noë and Thompson (2004), Madary (2013, 2017).
3 The distinction between organic and inorganic in this context is taken from Stiegler (1998).
4 An early emphasis on this theme is of course in Hegel's Phenomenology of Spirit (Hegel et al. 1976/2013). Dan Zahavi has done a great deal to develop this theme in classical Husserlian phenomenology (Husserl 1973; Zahavi 1996) and we also see it in major figures within the analytic tradition such as Davidson (2001), for example.
5 Ian Hacking (1997), another eminent philosopher of science, follows Popper in his turn to the objective basis of advanced cognition in writing technology. For some of the connections between this approach and Hegel's notion of objective mind, see Braver (2007).
6 The importance of the written word itself for the thought experiment is largely neglected in debates over the boundaries of the mind, but there are important exceptions. Richard Menary has explored the theme in a number of publications (his 2007, for example). Also see recent work by Regina Fabry (2020).
7 The preferences of the general public with regard to the use of XR may change as new applications are developed. For example, I suspect that immersive shopping may have wide appeal (Alcañiz et al. 2019).

References

Alcañiz, M., Enrique, B., & Guixeres, J. (2019). Virtual reality in marketing: A framework, review, and research agenda. *Frontiers in Psychology* 10: 1530. https://doi.org/10.3389/fpsyg.2019.01530.

Braver, L. (2007). *A thing of this world: A history of continental anti-realism*. Topics in Historical Philosophy. Evanston, IL: Northwestern University Press.

Brentano, F. (2015). *Psychology from an empirical standpoint.* Routledge Classics. Abingdon, Oxon: Routledge.

Carr, N.G. (2020). *The shallows: What the internet is doing to our brains.* New York: W.W. Norton & Company.

Chalmers, D.J. (2017). The virtual and the real. *Disputatio* 9 (46): 309–352. https://doi.org/10.1515/disp-2017–0009.

Clark, A., & Chalmers, D. (1998). The extended mind. *Analysis* 58 (1): 7–19.

Davidson, D. (2001). *Subjective, intersubjective, objective.* Oxford : New York: Clarendon Press; Oxford University Press.

Degenaar, J. (2014). Through the inverting glass: First-person observations on spatial vision and imagery. *Phenomenology and the Cognitive Sciences* 13 (2): 373–393. https://doi.org/10.1007/s11097-013-9305-3.

Derrida, J., & Spivak, G.C. (1974/2016). *Of grammatology.* Fortieth-Anniversary edition. Baltimore, MD: Johns Hopkins University Press.

Fabry, R.E. (2020). The cerebral, extra-cerebral bodily, and socio-cultural dimensions of enculturated arithmetical cognition. *Synthese* 197 (9): 3685–3720. https://doi.org/10.1007/s11229-019-02238-1.

Fodor, J.A. (1987). *Psychosemantics: The problem of meaning in the philosophy of mind.* Explorations in Cognitive Science. Cambridge, MA: MIT Press.

Forrester Consulting. (2021) The new way of working is immersive. *Forrester Opportunity Snapshot: A Custom Study Commissioned by Unity.* Retrieved from https://resources.unity.com/automotive-transportation-manufacturing-content/study-the-new-way-of-working-is-immersive

Freeman, W.J. (1960). Correlation of electrical activity of prepyriform cortex and behavior in cat. *Journal of Neurophysiology* 23 (2): 111–131. https://doi.org/10.1152/jn.1960.23.2.111.

Frith, C. (2005). The self in action: Lessons from delusions of control. *Consciousness and Cognition* 14 (4): 752–770. https://doi.org/10.1016/j.concog.2005.04.002.

Gallagher, S. (2006). *How the body shapes the mind.* Oxford: Oxford University Press.

Hacking, I. (1997). *Why does language matter to philosophy ?* Reprinted. Cambridge, MA: Cambridge University Press.

Hegel, G.W.F., Miller, A.V., & Findlay, J. (1976/2013). *Phenomenology of spirit.* Reprint. Oxford Paperbacks. Oxford: Oxford University Press.

Husserl, E. (1900/2001). *Logical investigations.* International Library of Philosophy. London; New York: Routledge

Husserl, E. (1973). *Zur phänomenologie der intersubjektivität: Zweiter teil: 1921– 1928,* ed. Kern Iso, Nijhoff, Den Haag, Netherlands.

Husserl, E. (1989). *Ideas pertaining to a pure phenomenology and to a phenomenological philosophy, Second book.* R. Rojcewicz, trans. The Hague, Boston, MA: M. Nijhoff ; Distributors for the U.S. and Canada, Kluwer Boston.

Husserl, E., & Carr, E. (1970/1984). *The crisis of European sciences and transcendental phenomenology: An introduction to phenomenological philosophy.* 6th pr. Studies in Phenomenology & Existential Philosophy. Evanston, IL: Northwestern University Press.

Hutchins, E. (1995/2006). *Cognition in the wild.* 8. pr. A Bradford Book. Cambridge, MA: MIT Press.

Illich, I., & Sanders, B. (1988). *ABC: The alphabetization of the popular mind.* London: Boyars.

Kelly, S.D. (2005). Seeing things in Merleau-Ponty. In Taylor Carman, and Mark B. N. Hansen (eds). *The Cambridge companion to Merleau-Ponty.* Cambridge Companions to Philosophy. Cambridge; New York: Cambridge University Press.

Kirsh, D. (1995). The intelligent use of space. *Artificial Intelligence* 73 (1–2): 31–68. https://doi.org/10.1016/0004-3702(94)00017-U.

Kittler, F. (2009). Towards an ontology of media. *Theory, Culture & Society* 26 (2–3): 23–31. https://doi.org/10.1177/0263276409103106.

Luria, A.R. (1976/1994). *Cognitive development: Its cultural and social foundations.* M. Cole (ed). Translated by Lynn Solotaroff and Martin Lopez-Morillas. 8. printing. Cambridge, MA: Harvard University Press.

Madary, M. (2012). Husserl on perceptual constancy. *European Journal of Philosophy* 20 (1): 145–165. https://doi.org/10.1111/j.1468-0378.2010.00405.x.

Madary, M. (2013). Placing area MT in context. *Journal of Consciousness Studies* 20 (5): 93–104.

Madary, M. (2017). *Visual phenomenology.* Cambridge, MA: MIT Press.

Madary, M. (2019). Husserl takes Santonin. In B. Glenney (ed). *The senses and the history of philosophy.* 1st edition. Rewriting the History of Philosophy. New York: Routledge.

Madary, M. (2022). The illusion of agency in human–computer interaction. *Neuroethics* 15 (1): 16. https://doi.org/10.1007/s12152-022-09491-1.

Madary, M., & Metzinger, T.K. (2016). Recommendations for good scientific practice and the consumers of VR-technology. *Frontiers in Robotics and AI* 3. https://doi.org/10.3389/frobt.2016.00003.

Menary, R. (2007). Writing as thinking. *Language Sciences* 29 (5): 621–632. https://doi.org/10.1016/j.langsci.2007.01.005.

Merleau-Ponty, M., & Smith, C. (1964/2006). *Phenomenology of perception: An introduction.* Repr. Routledge Classics. London: Routledge.

Millikan, R.G. (1987/2009). *Language, thought, and other biological categories: New foundations for realism.* A Bradford Book. Cambridge, MA: MIT Press.

Noë, A., & Thompson, E. (2004). Are there neural correlates of consciousness? *Journal of Consciousness Studies* 11 (1): 3–28.

Ong, W.J., & Hartley, J. (2012). *Orality and literacy: The technologizing of the word.* 30th anniversary ed.; 3rd edition. Orality and Literary. London ; New York: Routledge.

O'Regan, J.K., & Noë, A. (2001). A sensorimotor account of vision and visual consciousness. *Behavioral and Brain Sciences* 24 (5): 939–973. https://doi.org/10.1017/S0140525X01000115.

Popper, K.R. (1979). *Objective knowledge: An evolutionary approach.* Revised edition. Oxford: New York: Clarendon Press ; Oxford University Press.

Seth, A.K. (2013). Interoceptive inference, emotion, and the embodied self. *Trends in Cognitive Sciences* 17 (11): 565–573.

Siegel, S. (2010). *The contents of visual experience.* Philosophy of Mind Series. New York: Oxford University Press.

Slater, M. (2009a). Inducing illusory ownership of a virtual body. *Frontiers in Neuroscience* 3 (2): 214–220. https://doi.org/10.3389/neuro.01.029.2009.

Slater, M. (2009b). Place illusion and plausibility can lead to realistic behaviour in immersive virtual environments. *Philosophical Transactions of the Royal Society B: Biological Sciences* 364 (1535): 3549–357. https://doi.org/10.1098/rstb.2009.0138.

Sparrow, J. (2019). The car culture that's helping destroy the planet was by no means inevitable *Overland* 236.

Steinbock, A.J. (1995). *Home and beyond: Generative phenomenology after Husserl*. Northwestern University Studies in Phenomenology and Existential Philosophy. Evanston, IL: Northwestern University Press.

Sterelny, K. (2010). Minds: Extended or scaffolded? *Phenomenology and the Cognitive Sciences* 9 (4): 465–481. https://doi.org/10.1007/s11097-010-9174-y.

Stiegler, B. (1998). *Technics and time*. Meridian. Stanford, CA: Stanford University Press.

Turkle, S. (2017). *Alone together: Why we expect more from technology and less from each other*. 3rd edition, Revised trade paperback edition. New York: Basic Books.

Wehrle, M. (2015). Normality and normativity in experience. In Doyon, Maxime and Thiemo Breyer (eds). *Normativity in perception*. New Directions in Philosophy and Cognitive Science. Houndmills, Basingstoke Hampshire; New York: Palgrave Macmillan.

Zahavi, D. (1996). Husserl's intersubjective transformation of transcendental philosophy. *Journal of the British Society for Phenomenology* 27 (3): 228–245. https://doi.org/10.1080/00071773.1996.11007165.

Zuboff, S. (2020). *The age of surveillance capitalism: The fight for a human future at the new frontier of power*. First Trade Paperback edition. New York: PublicAffairs.

6 Extended Reality, Control, and Problems of the Self
An Ethical Analysis

Erick José Ramirez, Shelby Jennett, Dorian Clay, and Mohit Gandhi

Evolutionarily, we humans are social creatures. We evolved in the context of close-knit groups, and many of the cognitive, behavioral, and emotional mechanisms we take for granted today were adapted to fit those unique contexts. Because we're so interdependent, it's important for us to keep track of who we interact with, who is a reliable cooperator, and who has cheated us (Maguinness & Newell 2014). Evolutionary pressures have given us a suite of heuristics that make the process of identification and reidentification as seamless as seeing a familiar face or recognizing the sound of a friend's voice on the phone. Although the rapid technological changes wrought in the 20th century radically expanded the number of people we interact with, it's only recently that our technologies have threatened to fundamentally undermine our most basic epistemological heuristics for keeping track of others, and ourselves. This is a paper about those technologies and the problems they're likely to introduce in the not-too-distant future.

Extended (XR) technologies have only recently broken free from the realm of science fiction. Virtual reality (VR) devices like the Oculus Rift, Sony VR, and HTC Vive have gathered most of the major headlines since entering the market. In terms of readily available mass consumer devices, we haven't even spent a full decade figuring out what to do, and how to live, with these technologies. Today we are already seeing AR and VR technologies become increasingly woven into our everyday lives. For our purposes,

> [AR] is the use of technology in order to project digital content onto our experience of the physical world. Moreover, the content projected depends on (and adapts to) properties of the surrounding space.
>
> (Neely 2019)

Technologies in the XR spectrum include both AR devices (like the Microsoft Hololens) and VRheadsets. One of the primary features of XR is that

DOI: 10.4324/9781003359494-9

it empowers users to control their appearance in ways no prior technology allowed. Apps like Snapchat and Instagram offer users the ability to edit an image as it's captured, "in real time." Some filters apply simple color correction or film-grain effects, while others might overlay a 3D model onto the user's face. Some new personal electronics even have augmented features by default, such as Microsoft's Eye Contact, which adjusts photos so that users seem to be looking directly at the camera (Morton 2020). The range of filters present has led to a general increase in how consumers control their (digital) appearance. AR games like Niantic's PokemonGo!, which has generated more than $3.5 billion dollars in sales, have introduced AR to broad audiences (Desatoff 2020), and the wide availability of VR and AR applications has helped to charge public imagination over their uses.

As XR technologies mature, it's natural that users desire and gain increasing abilities to alter their appearances in XR spaces. We've always used the technologies available to us as a form of self-expression, and it's likely that XR technologies will follow this pattern. This paper explores the challenges that XR embodiment will introduce in our not-too-distant futures. We'll ask several questions about XR embodiment. First, we'll consider who should exercise control over XR embodiment overlays? If, as we'll argue, XR bodies can become as important to our sense of self as our physical bodies, we'll need a framework for making sense of who can rightfully exercise control over our XR bodies. We'll survey several candidates though all of them create ethical issues of their own.

Second, because XR embodiment allows for near-infinite modifications of the self, we'll need a framework for reliably tracking and identifying users across the metaverse. This becomes a special challenge when the traditional psychological and neurological tools we've evolved to use for this task are not available. In physical spaces, we evolved multimodal heuristics to keep track of our friends and loved ones. Unlike our physical bodies, however, XR bodies can be radically altered from meeting to meeting (even moment to moment), and so it's unlikely that our heuristics will serve us in the metaverse. Here, too, we'll survey potential options ranging from biomarkers, non-fungible token (NFTs), and patterns of our own social activities. What we need is a feature or tool that is both stick (that can pick out a specific individual) and stable (it can be used across several instances). The problem, we claim, will be in finding such a feature that avoids (if possible) the creation of large (state or corporate) surveillance mechanisms.

Thirdly, we note additional problems when interacting in social metaverse spaces that mix Human and Human-like (e.g., artificially intelligent) agents. Because XR embodiment allows for so much freedom of expression, we may not easily be able to distinguish other Human users from Human-like bots. To the degree that we'll want to distinguish between the two, we'll need a way to do so. Is it possible to distinguish between Human and Human-like bots in such spaces without resorting to potentially authoritarian or regressive responses? We're not sure, but we are certain that we'll need to address the ethical dimensions of these questions as we create, and begin to inhabit, metaverses.

The Extended Reality Spectrum

Before saying more about the issues that omnipresent XR embodiment introduces, we need to first explain not only what AR is but how it relates to a broader spectrum of "extended" reality (XR) technologies. Although we focus our arguments on AR, some of the issues we raise apply broadly to many forms of XR. We then discuss the ethical landscape of body modification to show that the reasons that make modifying physical bodies permissible (or impermissible) can transfer to the ethics of AR body modification. This will lead directly to the issues which are the focus for the rest of our chapter.

Augmented and virtual reality technologies include a diverse group of approaches, each generating simulated experiences for users. The differences between these technologies are not differences in kind but in the degree to which the creator of a simulated experience wishes to overlay simulated content onto the world. Both augmented and virtual reality technologies are part of the "extended reality" (XR) spectrum (Vi, Silva da Silva, & Muarer 2019). VR technologies aim to provide users with experiences whose contents, at least in principle, bare little to no relation to the physical world. AR is a way of referring to any technology wherein one or more features of the unaugmented world are overlaid with simulated content and where the unaugmented world serves as its own model (Stone 2000; Neely 2019). To better understand what we mean by having the world "overlaid with simulated content" while serving as its own model, consider the following examples.

In 2021, cosmetics giant L'Oréal announced a partnership with Modiface, a company that designs AR applications (Barras-Hill 2021). Modiface would be working to create AR makeup experiences for L'Oréal's customers. Users were able to either upload an image of their face or, in some cases, stand in front of a mirror where L'Oréal cosmetics could be digitally overlaid onto their faces. Modiface's AR cosmetic trial incorporates the features of AR we wish to highlight. The user's physical appearance serves as the base, the model, upon which simulated content (e.g., different hairstyles or cosmetics) are digitally overlaid. The model is overlaid so that the resulting image closely follows the user's facial structure. Although users can try on different hairstyles, cosmetic colors, and the like, they cannot radically alter all aspects of their underlying physical appearance (bone structure, etc.). In terms of an XR spectrum, AR cosmetic applications are fairly light: they modify a user's perceptions of the physical world in minor ways. In contrast to L'Oréal's AR cosmetic application, consider the following virtual reality application which more heavily modifies a user's perception of the world.

Unlike AR applications, which overlay content onto parts of the external world (e.g., a user's own face), VR games like Beat Saber create entirely simulated environments to place users in. Beat Saber is known as a "rhythm game" where players hit moving targets in sync with a soundtrack. It's the best-selling VR application to date having sold more than four million copies by the middle of 2021 (Lang 2021). The gamespace created by the developers of Beat Saber bears little-to-no relationship to the physical spaces players happen to be in (indeed the application is designed to transport users away from their surroundings and place them into the gameworld). VR applications like Beat Saber occupy a space distinct from contemporary AR applications.[1] Between AR cosmetics and VR rhythm games, we can imagine applications occupying a middle area. For example, imagine an application that seeks to entirely "reskin" one's experience of the world that goes further than L'Oréal's AR application in terms of overlaying content onto reality though does not go as far as traditional VR applications in terms of completely relocating the user into a new gameworld (i.e., an application that applies graphical overlays to replace all aspects of a user's physical surroundings to make the world appear as if it were a user's favorite animated universe).

We now focus largely on AR applications (i.e., applications that overlay information, sounds, sights, and so on onto features of a user's physical environment). In particular, we focus on applications that allow users to overlay physical, visual, and auditory content over their bodies. We'll refer to applications that allow users to have this kind of existence as allowing users to experience "XR embodiment."[2] Before exploring the ethics of XR embodiment, it's important for us to examine the (rough) consensus emerging regarding the ethics of modifying one's physical body. We'll argue that it may appear natural to extend ethical principles governing physical body modification to AR body modification. Such a move, however natural, creates distinct ethical issues which carry epistemological, ethical, and social costs.

Autonomy, Representation, and Body Modification

Perhaps the least surprising aspect of XR embodiment is that individuals will use the technology to alter their appearance. From makeup to exercise to fashion and cosmetic surgery, people have availed themselves of the technologies available in order to modify or express themselves. Such alterations are carried out for many reasons: in response to aesthetic or gender norms, to bring physical presentations in line with a person's sense of self, or as radical performative aesthetic acts of protest or affirmation. XR embodiment and body modification is, in these ways, an extension of existing practices.

Although modifications to the body always include an ethical dimension (McCabe, de Waal, & Fabri 2017; Sarajlic 2020), we argue that XR embodiment and body modification introduce ethical challenges to the ethics of body modification. We, here, survey the ethical issues surrounding body modification. We do this for two reasons. First, it will seem natural to extend the ethics of body modification to the ethics of XR embodiment, and we survey the scene to see what a resulting ethics may look like. Second, despite the naturalness of this extension, the resulting ethics would be incomplete. The ethics of physical body modification won't fully account for the special nature of XR embodiment including its range of expression and malleability.

When assessing the ethics of body modification, bioethicists traditionally focus on three considerations. These include competence (of the agent requesting the modification or procedure), non-maleficence (i.e., is the modification or procedure in the best interest of the person?), and considerations of justice (i.e., given scarce medical resources, is satisfying this request justifiable?) (Gillon 2015). Competence is the most basic of these. To deem a person's desire to modify their body permissible, they must be judged competent to make and act upon their desires. Competence, however, is a complex series of capacities, and the degree of competence we seek may also vary depending upon the severity of the modification (i.e., children, to some degree, may choose what clothes to wear, but parents, and in some cases third parties, usually must make and approve of choices with longer lasting or permanent effects on a child's life).[3]

People who want to permanently modify their bodies must be both competent *and* rational. Individuals must be able to understand and appreciate their decision and its outcomes (Culver & Gert 2004). Understanding requires that a person possesses the capacity to comprehend the general nature of the procedure and its consequences (Culver & Gert 2004). For example, if an individual wants to get a tattoo, they must understand that, in general, tattoos carry some risk of infection and viral transmission risk if the instruments are not properly cleaned, etc. To appreciate such consequences means that a person realizes that these risks may happen *in their specific case*.[4] The competence requirement is often paired with some view of minimal rationality (i.e., a capacity for drawing intersubjectively reasonable inferences).[5]

Consider a traditionally unproblematic case of cosmetic body modification. Jasper wants calf implants. Jasper understands the nature of the procedure and appreciates its risks. Their reasons for seeking the implants are aesthetic in an intersubjectively reasonable sense (i.e., others who don't share Jasper's values can understand their desire even if they personally would not elect to get the implants themselves). Jasper is well aware that they will not gain any actual muscle mass, only the appearance of it. In such a case, Jasper's desire, whatever we may personally think of it, is morally permissible for them to act upon so long as the medical risks involved are reasonable.

As a contrast, consider a case of someone diagnosed with what is known as body dysmorphic disorder (BDD) (American Psychiatric Association 2022). BDD is characterized, in part, by an irrational, or delusional, understanding of one's own body (especially a particular region of the body that is the focus of a BDD). Patients seeking cosmetic body modification, especially those who seek multiple alterations, are more likely to be diagnosed with BDD and more at risk (Higgins & Wysong 2018). Surgeons who suspect that a client is seeking body modification *as a result* of BDD are cautioned against performing these surgeries. This is because the delusional nature of the disorder compromises a person's ability to understand and appreciate the outcomes of modification of their bodies.

With respect to embodiment in XR spaces, it's natural to extend existing ethical principles to new cases. Despite this naturalness, such an extension fails to meet the challenges posed by XR embodiment. Unlike traditional body modifications, XR embodiment allows for near limitless and reversible changes to the body including those that go beyond modifications of physical appearance. While we can fork a tongue, insert (to some degree) a range of implants, and cover our bodies with tattoos, the physical body admits of only so much modification. With XR bodies, this is not true. Since modifications are limitless, the degree of freedom afforded by XR embodiment far exceeds the freedom currently available using cosmetic and surgical forms of embodiment. As we shall see, for these and other reasons, the ethics of XR embodiment cannot easily be captured by existing principles governing body modification.

Issue 1: Control, Representation, and Identity

While XR embodiment gives users immense new power over social presentation, it also poses unique challenges. The ethics of body modification, concerned with individual autonomy, welfare, and rationality, must be modified to make sense of the psychological and social consequences of XR embodiment. In this section, we ask who should be empowered to control how individuals are represented in AR spaces. We survey four responses to this question and show how each creates problems of its own. We consider the ethical implications of empowering individual users to control how they look, sound, and feel to others. Additionally, we explore the consequences of empowering users to control how *others* are embodied. Because these options only make sense against a backdrop of enforcement, we consider the consequences of empowering corporations or states to regulate XR embodiment.

XR embodiment can have powerful effects on users. In a study on XR embodiment using VR, Guo Freeman and her colleagues (2020) investigated the psychological effects of avatar creation and embodiment. They found that not only did subjects feel "more engaging, intimate, and personal with their avatars in social VR in contrast to traditional virtual worlds and online games" (4) but they felt a sense of ownership over their bodies. For example, two trans women participants described the feelings they experienced while embodying female avatars:

> Using a feminine avatar makes me confident not only in VR but also in real life. I feel like that would be actually more real than the real you in real life. Because in real life, you're stuck with what you were born with. But in VR, you can be what you truly feel like you are inside. This experience actually gave me confidence to start my [gender confirmation] procedure in the real life.
>
> (5)

> [B]y using a feminine female avatar, I found that I was just more comfortable with that body, and it's kind of what I learned about my identity. That was the evidence to myself to consider which direction I wanted to take my actual body outside of the VR. If I found I was happy in VR about my body and I was not happy with my body outside of the VR, why not change it?
>
> (5)

In some cases, in other words, XR bodies can lead to an affirmation or reappraisal of a person's identity and our relationship with our physical bodies. It's important to distinguish the effects of XR embodiment from other (i.e., gaming) simulations where users may be said to be embodied. Simulations where all users are embodied in the same way (because they take on the role of the same hero character) are unlikely to trigger the sorts of experiences Freeman's subjects are describing. A user's ability to design their XR bodies is essential.

Freeman's research suggests both that our physical embodiment can affect how we create and shape AR bodies but also that how we are embodied in AR can cause us to rethink our relationships to our physical bodies. These effects can be strong enough to induce clinically significant dysphoria. "Snapchat dysmorphia" refers to clinically significant distress caused by a mismatch between a person's appearance as viewed through digital filters (i.e., XR embodiment) and a person's perceptions of their unfiltered physical self (Kleemans et al. 2018; Ramphul & Mejias 2018; Rice, Graber, & Kourosh 2020; Tremblay, Tremblay & Poirier 2021). It's for this reason that our norms, policies, and laws regarding XR embodiment deserve special attention.

AR spaces shared by multiple agents require rules or regulations governing how entities in those spaces can look, sound, feel, etc. both to themselves and to one another. Who should have control over how users appear in such spaces? In the following sections, we survey four responses to this question. The upshot of our analysis is showing that current norms, laws, and intuitions regarding physical body modification are insufficient to address the ethical issues introduced by XR embodiment.

Perhaps the most natural response to our question is that individual users should control their own appearance. This aligns closely with current intuitions about fashion, physical body modifications, and body presentation generally. People are typically permitted to present themselves in whatever way they wish.[6] The value of individual autonomy and self-expression are overriding, in most contexts, in the face of offense to others occurring from aesthetic difference. There are exceptions, as some contexts enforce a dress code, but for the most part, public spaces are areas where one's appearance is grounded in individual autonomy.[7] Despite the power of this intuition, and its connection with the ethics of physical body modification, there are risks associated with giving users complete control over their XR embodiment. XR embodiment, unlike physical embodiment, allows for greater variation in both kind and degree of body modification. While physical body modifications are always limited by the body itself (e.g., there is only so far one can biologically modify the Human organism), no such limits exist for AR body modifications. Because XR embodiment offers nearly limitless possibilities for self-expression and modification, it also offers near-limitless avenues for bad actors to cause harm.

Bad actors present challenges to the widespread adoption of new technologies, and society's introduction to social XR is no exception. One type of bad actor has come to prominence over the last 25 years: the Internet Troll. Trolls are people who engage in trolling. "Trolling involves deliberate, deceptive and mischievous attempts to provoke reactions from other online users" (Golf-Papez & Veer 2017). Unsurprisingly, researchers have come to the conclusion that "anonymity leads to more trolling" (Guo & Caine 2021). How might trolls take advantage of XR embodiment? Opportunities abound. We can imagine trolls taking on the appearance of a loved one, perhaps aided by deepfake technologies (Westerlund 2019), to engage in inappropriate conduct. Trolls may embody themselves as their target's harasser or abuser in order to traumatize. Trolls may don a form of digital Blackface with the intent of perpetuating racist stereotypes or engage in racist harassment. Given the wide-ranging powers of XR embodiment, a policy that empowered users, without limitation, to embody themselves however they wish has significant downsides. The negative effects of a norm that unconditionally empowers users to embody themselves however they wish should lead us to search for an alternative.

One alternative would be to allow users to control how *others* appear to them in social AR spaces. Note that this alternative deviates from the norms in many societies covering physical embodiment. Individuals are normally free to choose how they dress, behave, talk, and modify their bodies in public spaces so long as they don't directly harm others. While some may find the behavior or appearance of others offensive, such offense is not typically thought to license regulation in most cases.[8] In physical spaces, while we are free to avoid engaging with others, we may find annoying or offensive; we're not typically allowed to interfere with another's ability to represent or embody themselves. However, empowering users to control how others present themselves in social AR worlds would address many of the issues laid out earlier. Trolls lose the ability to harass by embodying themselves in offensive, racist, or hurtful personas. However, allowing users to block AR overlays they find annoying or offensive is both disempowering (to those who want to embody themselves in those ways) and a form of digital erasure (users may literally be made invisible or silent). In private, physical, and virtual spaces, this may be an acceptable tradeoff, but in a metaverse that acts as a digital public square, the cost is significant.

Imagine, for example, a user is morally offended by LGBTQ relationships. Such a user would gain a power they wouldn't have in physical spaces: the power to erase LGBTQ-identifying users from their metaverses. The kind of social metaverse spaces we're targeting, one where users carry out economic, political, and educational activities common in physical spaces, is being developed now. Mark Zuckerberg, CEO of Meta (a corporation investing most heavily in the development and creation of metaverse spaces) claimed that "it's going to take a while for it to get to the scale of several hundreds of millions or even billions of people in the metaverse... that's the north star. I think we will get there" (Novet 2022). Social AR spaces like these, which seek to mimic public physical spaces, deserve to be scrutinized especially seriously.

Luciano Floridi argued that digital spaces

> Where humanity spends more and more time and where more and more activities take place directly or indirectly, from education to work, from socialisation to entertainment, from commerce to finance, from the exercise of justice to political discussion, from research to journalism

merit a distinct kind of regulatory framework (Floridi 2021). This kind of space

> Influences every other space, even the physical one...It is a space that should be conceptualised and governed more like a condominium... rather than like a new frontier that can be appropriated and colonised by anybody, or like a space that belongs to no one, like the Moon.
>
> (Floridi 2021)

Seen from that perspective, it's harmful to remove an individual's power to shape their appearance or express their values (however noxious others may find them).[9] While empowering users to control XR embodiment for others solves some of the issues stemming from XR embodiment, it leaves others untouched. Deepfakes again emerge as a problem (de Ruiter 2021; Johnson & Diakopoulos 2021). In addition to digitally censoring or erasing the XR overlays of others, users empowered to control the XR embodiment of other users can use this power to overlay Deepfaked nude images of users for their own amusement or to harass others. Problems like these suggest that a greater regulatory role needs to be played with XR embodiment and that leaving things simply in the hands of users isn't an adequate way to resolve the ethical issues inherent in XR embodiment.

Floridi argues that shared AR spaces should be regulated like condominiums, what might that mean? We explore three options. One approach is to empower the corporations that build and maintain these worlds to regulate them. A second approach empowers states to regulate XR embodiment for their citizens (or corporations located within their borders). A final approach takes the condominium analogy more seriously in the sense that user-created communities that function like homeowners associations (HOAs) (typically composed of individuals within a specific community) control embodiment in their spaces. Members of the community democratically elect residents of the community to create an HOA and are empowered to set regulations for them including levying fines, collecting funds, and setting standards for decency.

Currently, corporations are given wide leeway when it comes to regulating their users' behavior and speech.[10] Individuals must acknowledge and consent to abide by Terms of Service (TOS) agreements before corporations allow them access to their platforms. Each TOS defines standards for decency, harassment, and privacy. While TOS agreements may resemble one another across platforms, each platform may write its own terms for its users. When it comes to regulating XR embodiment, this corporate approach extends existing practices from one domain (the Internet) into another (social XR). This approach would seem natural to users who already agree to dozens of such contracts. Despite this naturalness, problems emerge. XR technologies are more closely tied to one's sense of self given the interdependence between physical and XR embodiment. While users may be comfortable empowering corporations to regulate some aspects of their speech, those same users may balk at giving corporations control of their ability to embody themselves. Additionally, there are concerns about corporate motivation to fairly apply its TOS to users. The imperative to maximize shareholder value drives social XR spaces. Historically, this has led corporations to regulate popular members (those that drive engagement or the metaverse equivalent to "likes" or "follows") with a lighter touch than others. If the metaverse is to have the social, political, and economic importance its proponents claim, then access to these spaces should be regulated fairly. Allowing corporations, motivated to maximize shareholder value, to regulate users is likely to be inconsistent with procedural fairness.

Democratic state control of XR embodiment offers some advantages. Ideally, democratic states operate under principles different from profit-driven institutions. They're committed to procedural fairness in the treatment of their citizens. In practice, states fail to uphold these values though failures of procedural fairness generate protest, and liberal democratic states have mechanisms built into them to allow for changes to unfair laws. States are also already empowered to regulate physical embodiment: they set laws determining decency, and they regulate when, why, and how modifications to the physical body are made. This gives state approaches some prima facie plausibility.

Democratic states ought to treat XR embodiment more fairly than corporations. Like corporate approaches, however, this plausibility comes at a price. There are wide-ranging disagreements about how states should regulate physical embodiment (e.g., laws regulating abortion, gender confirmation, and cosmetic surgeries are matters of disagreement). Concerns about tyrannical majorities or minorities wielding power over bodies is a fact of life when it comes to physical embodiment, and these concerns carry over to XR embodiment. In addition, states empowered to regulate XR embodiment are simultaneously granted the power to digitally exile citizens even in physical spaces. If social XR spaces are as ubiquitous and important to our social, economic, and political lives as its advocates claim they will be, then states with the power to regulate XR embodiment can choose to bar access to these overlays, or use their power over XR embodiment to overlay content onto our bodies. This would not only take away an intimate form of self-expression, it could also render a person unrecognizable to others. A "blank" overlay could be used to punish dissidents by making them highly visible but impossible to see or hear. Similarly, the power to see others can be removed by applying anonymous overlays onto citizens as a form of AR ostracization. Caution must be exercised if we choose to empower in this way.

A third alternative takes literally Floridi's advice to regulate digital spaces as we regulate condominiums. Call this the HOA model. Unlike corporations, HOAs don't have capitalist incentives baked into their structure and may be willing to enact regulations that don't maximize their bottom line but which make the community better (e.g., collecting fees is a financial hit that HOA members take on to keep their communities maintained). Because online communities and spaces are often transnational, these metaverse spaces can be composed of individuals from around the world who agree to forge their own social contracts. Their regulatory powers are, in some ways, broader than those of a state which are limited by geographical borders. HOA governance can be global but also representatively democratic. These features make the HOA approach attractive. On the other hand, communities empowered to regulate themselves may choose to organize around toxic norms.[11] Racist, misogynist, or terroristic HOA communities can pose genuine threats to other communities and can embrace norms that would be voted down by citizens of states. The HOA approach imagines the metaverse and XR embodiment as a series of walled-off communities governed by different norms. This approach, while it has many advantages, would make it difficult for individuals to engage with one another in truly public ways (such public spaces may indeed be impossible on this model without decomposing into state or corporate approaches).

XR embodiment will become increasingly important to us as our social, economic, and political activities shift to the metaverse. A ubiquitous social metaverse requires the existence of norms, rules, or regulations to govern who has control over XR embodiment and what limits such embodiment can take. The options we've explored are all plausible (alone or in combination), but each requires tradeoffs between values (autonomy, freedom of expression, freedom from harm, and the regulation of offense/hate speech). Even if we are able to come to consensus about how to regulate XR embodiment, other problems of the self persist.

Issue 2: Tracking the Self Across the Metaverse

Suppose we come to a conclusion about how to control XR embodiment. Users in this future exist across private, semi-private, and public social XR spaces where they're likely to want to encounter and re-encounter friends, enemies, and meet strangers. This basic, but important, feature of social environments leads to a new problem. "The accurate perception of others is a fundamental aspect of social cognition, allowing us to detect the intention, attention, and identity of an individual (among other attributes)" (Maguinness & Newell 2014), and this would be as important in XR as it is in physical spaces.

We've evolved complex multimodal psychological capacities that help us to keep track of ourselves and others in physical spaces. When we recognize old friends, enemies, or our colleagues, we do so by picking up, often unconsciously, on their physical features. A friend's face (Kanwisher, McDermott, & Chun 1997; Sugiura et al. 2005; Maguinness & Newell 2014), the sound of their voice (Kaplan et al. 2008), the way they move (Cutting & Kozlowski 1977), even the way they smell (Milinski et al. 2013) can serve to verify they are who we think they are. Metaverse spaces rob us of these heuristic tools. XR embodiment makes it possible for someone to look, sound, and move very differently than they do in physical spaces. This introduces a new problem for inhabitants of social XR spaces. Because these spaces bypass the heuristics we normally use to keep track of others, how do we know who others are? Some form of identity verification is necessary, but is it possible to verify someone's identity while simultaneously resisting the pull of totalitarian corporate or state control over XR embodiment?[12] Our options are less clear than they were about who should control XR embodiment; nevertheless, in this section, we survey three strategies for fixing identity in the metaverse.

One approach to fixing identity across metaverse spaces is to rely on biomarkers. Fingerprints, facial structure, iris scans, and the like function as uniquely identifying markers that individuals can use when they enter the metaverse in order to verify that the XR-embodied individual claiming to be the user is, in fact, the user. Such an approach has some obvious advantages: many today already rely on biomarkers to gain access to their smartphones and computers because biomarkers are strongly tied to specific users. Using a fingerprint to unlock a phone relies on the same logic. If only *you* should have access to your phone (XR body), then we can limit access by relying on a biomarker that uniquely tracks you. While it is a natural response to the problem of fixing identity across metaverse spaces, the use of biomarkers can serve to deepen ongoing social and metaphysical controversy about the self and its relation to the physical body.

The intuition that individuals can, and should, be empowered to alter their physical bodies in line with their subjective sense(s) of self-identity is an intuition whose popular plausibility has changed over the last fifty years. Transhumanists, for example, argue that the self and the biological body are often separable and that we have a moral obligation to use our technological powers to augment or modify our underlying biology because it's not essential to who or what we are (Bostrom 2005; Vita-More 2021).[13] The same intuition is shared, broadly speaking, by other trans-movements including individuals who identify as transracial (Haslanger 2000; Tuvel 2017) or transabled (Davis 2012).[14] Our mentioning of these trans-movements is not intended to validate (or invalidate) these identities but, instead, to note that questions about the relationship between the self and the natal or unenhanced body are very much in contention today, and the social resolution of these debates into the future is hard to predict. Whether someone claiming to experience body integrity identity disorder (BIID) represents a socially valid transabled identity, or whether transracial identities are socially or metaphysically legitimate are issues whose long-term resolutions are unclear at present and even harder to predict in the future. For us, it's enough to say that social ideas about the relation between the self and the unaugmented or unmodified natal body are apt to continue changing and that it would be advantageous if our analysis of XR embodiment were sensitive to these changes. Although biomarkers would help settle questions about metaverse epistemology, they would do so by imposing a set of norms that may appear regressive to those inhabiting these places in the future. Additionally, biomarker approaches introduce massive data security risk. Corporations, or states, who would become responsible for verifying identities, would need to make biomarker data available across a variety of virtual spaces in order for biomarkers to help us track identity across the metaverse. Given the intimate nature of this data and the history of data security breaches, introducing such a risk is unattractive.

Although biomarkers may ultimately be the best approach for solving the problem of identity verification, it would be good to find alternative methods of fixing identity that don't rely on them if only to avoid relying on problematic assumptions about biomarkers and the self. One alternative to biomarkers for fixing identity in the metaverse is to rely on hashing (encrypting) user identities by using non-fungible token technology as a way of storing a record of a user's identity as a part of a global distributed ledger known as a blockchain. While the technical details behind blockchain technology are important (Schmeiss, Hoelzle, & Tech 2019; Dowling 2021), their philosophical value lies in their potential to keep track of identities in a way that

1 Is independent of state or corporate monitoring (blockchain integrity is maintained through global distributed peer-to-peer networks)
2 Are highly resistant to tampering
3 Allow users the freedom to choose what to use to serve as their private key to link them to their NFT identity (a photo, a favored poem, etc.)

Blockchain technology is famously associated with digital cryptocurrencies like Bitcoin though NFTs are also rapidly becoming the subject of widespread interest and investment in the artworld (Moscufo 2021). In much the same way as a user can access their Bitcoin wallets globally and without government oversight, an NFT approach to identity would allow individuals to link their XR bodies to a unique key that they can take with them throughout the metaverse and that can serve to guarantee that the individual in possession of the key can access the relevant XR body.

Because NFTs are unique tokens (they point to a specific transaction in the blockchain at a specific time,) they are *stable* ways of marking identity. Because they don't rely on biomarkers, the NFT approach to identity serves as an alternative solution to tracking identity. Lastly, because NFTs are tracked on global distributed ledgers, they avoid problems associated with state or corporate control of intimate information. Despite these attractive elements, the NFT model has several drawbacks. First, although NFTs are *stable* markers for identity (they are robustly secure and unique), they are not especially *sticky*. When the artist known as Beeple (a.ka. Mike Winkelmann) sold his NFT artwork "Everydays" for $69.3 million to an anonymous buyer, what was sold was Beeple's key to a specific transaction within a Blockchain that represents the unique NFT known as "Everydays" (Moscufo 2021). In much the same way as someone might partition a Bitcoin and use it to purchase something, NFT markers for identity can be traded between users via the trading of private keys that give access to the NFT itself. Because these trades can be anonymous, NFTs make for poor guarantors of individual authenticity (i.e., you may be interacting with someone embodied in exactly the XR body you previously interacted with, but this XR body may now be accessed by a different person who has been given, or stolen, the relevant key). NFT identity allows for two serious problems: a user may have access to multiple identities, and some users may permanently lose access to their identity by losing their key. It may be possible to make NFTs stickier, but this may, counterintuitively, make them less attractive as an alternative to biomarkers (e.g., by requiring that users register their keys in a central identity repository that can, itself, become compromised).

The NFT approach to fixing identity was attractive because it promised to tie identity to something other than a biomarker thus avoiding the assumption that one's true identity is or ought to be tied to the physical body. It also didn't require that a state or corporation create a global repository of biomarkers to do its job. A third approach to tracking XR embodiment in the metaverse promises to preserve these aspects of the NFT approach. This approach, the "identity as activity" (IAA) approach, fixes user identity by tracking the patterns of connections, activities, and relationships that users have. The IAA approach is grounded on existing programs to deanonymize anonymous online activity by tracking a user's patterns (Su et al. 2017). Although the internet is vast, we tend to only regularly visit a very small set of sites and directly engage with a limited number of people despite having dozens, or even thousands, of followers on social media networks. These patterns of activity can serve to identify us uniquely even if the data itself is initially anonymized. This fact about our activity makes it possible to correlate specific patterns of activity with specific users fairly accurately. Initial studies have correctly tied over 70% of anonymous users back to their personal social media accounts even without complete information of their online activities (Su et al. 2017). Our freely chosen activities can reliably track us throughout the metaverse. Another alternative would identify individuals on the basis of their literal activity. Anonymous information about how users move through physical and virtual spaces (collected from handsets and headsets) can be used to successfully reidentify them (Falk et al. 2021).

While the IAA approach might seem dystopian, it lends itself to the task of identity tracking across the metaverse. Research suggests that our patterns of activity, at any given time, are both sticky and stable. Unlike the NFT approach, our patterns of activity cannot be traded. Unlike the biomarker approach, IAA relies on tying users to something less problematically connected with their identities (i.e., it's less controversial to say we are the sum of our activities, friendships, interests, or movements than it is to say we *are* our physical bodies). In this sense, IAA has advantages over earlier views. It does, however, have its own drawbacks. First, we may have doubts about the lasting *stability* of our patterns. While our activities, friendships, and literal physical movement patterns may be stable at any given point, our patterns over longer periods of time are not. Thinking back to the friends, websites, games, and social media activities from our teenage years should suffice to show that our modes of engagement change over time (though perhaps they change more slowly and less dramatically as we age). While IAA may be able to identify us in the short term, it may have trouble keeping that identity stable in the long term without turning over access to the history of our metaverse activities.

Another concern about stability is more intentional. Users specifically seeking to avoid having their activities tracked may cultivate distinctly different patterns of activities (i.e., the equivalent of burner accounts) to mask their identity. Users may distinctly avoid their favored sites, friends, links, and adopt new ways of moving about while they engage in activities they don't want associated with their accounts. This leaves open the possibility that IAA allows for deception and a breakdown of stability of identity across the metaverse.[15] Lastly, IAA would further diminish the information over which we can claim digital privacy in a way that runs counter to movements already in place in many countries. For example, the enactments of the general data protection regulation (GDPR) in EU countries and the California consumer privacy act (CCPA) suggest a shift toward greater appreciation of, and protection for, privacy especially when it pertains to sensitive or intimate information about ourselves. Because the IAA requires that we provide unfettered access (to corporations or states) to our online and metaverse activities, interactions, and media engagement, such access runs counter to these movements (we leave it to readers to determine whether these problems suffice to rule out IAA).

Regardless of how we decide who has *control* over XR embodiment, we need a way to ensure that we can *trust* that the individuals we interact with in the metaverse are the people we think they are. Biomarkers, NFT identity hashing, and our own patterns of activity are, to varying degrees, up to the task though not without causing problems of their own. On the assumption, we can address these two issues; we're still not ready to dive into the metaverse. XR embodiment creates another problem that we'll need to be ready to address. Human-like bots, agents who are intentionally or unintentionally designed to naturalistically interact with people, are already making their way into our lives. It's likely that the metaverse will be populated by both Human and Human-like agents. Given the nature of XR embodiment, it will be hard for us to be sure that we're interacting with humans and, for some of us, this might matter.

Issue 3: Human-Like Bots

XR embodiment allows users to tailor how they look, sound, move, and feel in order to best represent themselves. Unlike physical embodiment, XR embodiment is limited only by our imagination and hardware. In the unaugmented world, physical embodiment limits a Human-like bot's ability to convincingly pass as Human (at least up close). Human-like bots in the metaverse won't have this issue. AR-embodied Human-like bots can not only look, sound, move, and (eventually) act like Human users, depending on how we resolve the issue over identity tracking, they may also have NFT keys or their own identity-enabling patterns of activity. Is it possible to successfully distinguish Human users from non-Human users in the metaverse?

Our use of "Human-like" picks out those agents whose design, intentionally or unintentionally, allows them to mimic Human behavior, including emotional expression, in a way that can deceive Human users (Weber-Guskar 2021). Human-like bots also already exist. Google's "Duplex," is a digital assistant designed to mimic human voice and inflection patterns to naturally interact with Humans while carrying out its owners' ends (i.e., make appointments, schedule meetings, etc.) (O'Leary 2019). Ethical issues have already been raised about the use of Human-like bots revolving around "meeting user needs, the ethics of deception, and reinforcing social stereotypes through conversational agents" (Pradhan & Lazar 2021). Human users usually want to know whether they are engaging with another person or with a Human-like bot and bots are often designed not to provide clear answers if asked whether they are Human (Gros, Li, & Yu 2021). We don't directly address the ethics of designing or deploying Human-like bots. Instead, we assume that interactions with Human-like bots, both because of their effectiveness and scalability, are likely to be common in the metaverse. Given ethical concerns about deception and stereotypes along with our preferences for interacting with other Humans (Baron 2013), is it possible to successfully demarcate Human agents from Human-like bots in the metaverse given the special nature of XR embodiment?

Of the issues we raise in this chapter, this is the most abstract though it gets at the heart of questions about the value of social interactions. The answer to this question is connected to prior issues and may seem obvious: states (or corporations) can legislate (or regulate) that Human-like bots are distinguished from Human denizens of the metaverse. This requires ceding power over XR embodiment to state or corporate structures but may be a reasonable solution. However, given the problems raised for these options earlier in the chapter (that the metaverse is global and state regulatory power is not, that corporations are not concerned with equity or equality, etc.), it's worth maintaining some skepticism about how successful these responses may be. Additionally, empowering states or corporations to regulate XR embodiment doesn't yet respond to the central question we're exploring: how would states or corporations *know* whether the metaverse agent in question is a person or a Human-like bot? Human-like bots can access NFT keys and have patterns of engagement and interaction that are uniquely their own. Although we argued that relying on biomarkers to track identity is problematic, biomarkers can be useful to distinguish Human from Human-like while avoiding the issues raised earlier.

A central concern about using biomarkers to track identity was that using sticky and stable biomarkers only made sense if we assume that the unenhanced/natal/biological body was representative of one's "authentic" self and that XR embodiment was not important in the same way. Given the social, political, ethical, and metaphysical challenges underway with respect to the significance of the body to identity, such an assumption may be regressive. When it comes to the distinction between Humans and Human-like bots, however, we need not rely on biomarkers like those (fingerprints, iris scans, facial structure, etc.). If we're only interested in distinguishing Humans from Human-like agents, then any biomarker capable of marking this distinction is good enough. The presence of a heart rate, blood pressure, a pulse, etc., wouldn't tie a biomarker to someone's specific identity (thus avoiding the assumption that biomarkers are essential to who we are *as people*) but would be capable of distinguishing biological agents from non-biological bot agents.[16][17]

Some may worry that any proposal that marks a rigid distinction between Human and non-Human agents may seem morally dubious due to the long history of denying personhood to members of our own species for morally irrelevant reasons. We leave open the possibility that Human-like bots may eventually be agents in their own right, fully deserving of moral attention. In such a future, it may be worth revisiting the intuition that Human-Human interactions are morally preferable, or that our treatment of Human-like agents is morally benign. Additionally, given what we've already said about the nature of XR embodiment and the fact that some are already willing to invest great value in their non-biological AR bodies, it's not clear that the distinction between biological Humans and Human-like bots is metaphysically defensible.[18]

Conclusion

In less than 20 years, Internet technologies fundamentally altered how we live, work, socialize, and engage politically. Metaverse technologies are only now becoming widely commercially available though they have yet to be incorporated into the basic structures of society. In this chapter, we examined the nature, and ethics, of XR embodiment in metaverse spaces. XR embodiment promises to change how we think about ourselves, the people around us, and has at least as much transformative potential as Internet technologies. In order to prepare for its transformative power, we need to be ready to meet the challenges XR embodiment will introduce. Specifically, we need to create a framework for deciding how XR embodiment should be controlled and understand the moral tradeoffs that any response to this question imposes. We need to develop better frameworks to understand the epistemology of identity in the metaverse so that we trust that our interactions with people in the metaverse are with the people we think they're with. Biomarkers may be necessary to do this but come with costs of their own. Lastly, we need to think, and perhaps rethink, our desire to differentiate Human agents in the metaverse from Human-like bots. XR embodiment can make it impossible for agents within the metaverse to make the Human/non-Human distinction, and we need to be ready to critically examine the importance of the distinction itself.

Notes

1 There are two exceptions worth noting. Many VR systems allow users to create a gaming area where they can safely move around by tracing out space in the physical world and can alert users when they're nearing the edge of these

safe spaces in order to avoid accidents. Such spaces would then bear an important relationship to real world spaces. Secondly, we've been discussing the XR spectrum almost entirely in terms of two modalities: visual and auditory. It's certainly possible, and indeed likely, that the XR spectrum can incorporate other modalities as well (haptic feedback, olfactory and gustatory stimuli, etc) though applications that utilize other modalities are currently rare.

2　Embodiment may, ultimately, not be the best metaphor to express this concept. In the final chapter of this volume, we use the term "body blending" to describe a similar experience. What matters, for our purposes, is that the user's own sense of who and what they are can come to essentially include (even prioritize) digital bodies in addition to (or even instead of) physical ones.

3　Though see Steensma and Cohen-Kettenis (2018) for one example of when and how the issue of respect for a child's autonomy and competence can become more complex.

4　There is an interesting debate about how strongly these elements must be interpreted. For example, some scholars argue that it's impossible for us to understand and appreciate medical interventions that result in what are called "transformative" experiences (which alter the user's sense of self so strongly that the pre-intervention person cannot successfully imagine, and thus cannot truly appreciate, what it would be like to be their post-intervention self) (Paul 2016). Additionally, reaching the conclusion that a person's *desire* to modify their body is grounded in autonomy/competence need not automatically make that procedure justifiable. Other ethical variables may speak against it (i.e., a tattoo artist may find it morally objectionable to tattoo a swastika on a potential client despite their client understanding and appreciating the risks of the procedure and the meaning of the image).

5　Fischer and Ravizza (1998) argue for a conception of rationality they refer to as "moderate reasons responsiveness" that relies on this conception of intersubjectively valid inference. Elsewhere, we've argued that "To say that rationality is intersubjective is to say that it requires third parties to *understand* a person's reasons in order for those reasons to be rational. For a desire to be understood as rational, it must be comprehensible to at least some third party communities" (Nelson & Ramirez 2017). Importantly, understanding another's chain of reasoning is consistent with *disagreeing* with that pattern. One need not be religious to understand someone's religiously motivated desires about the afterlife. The existence of a community of like-minded individuals is additional evidence in favor of a pattern of reasoning being "intersubjectively valid."

6　Though here too we pause to note that these choices are typically constrained by legal frameworks. Ultimately, legal or policy-based enforcement will need to be a component of any response to issues about XR embodiment and control. We distinguish those issues here in order to make clear that the issues associated with each option are distinct from one another and that stressing one response (user freedom) over another (corporate regulation) will result in differing tradeoffs.

7　We'll say more about the role of corporate or legal enforcement in just a moment.

8　Limit cases exist (i.e., laws governing nudity or obscenity), and we can imagine applying similar limit cases to social XR spaces as well. The alternative regulatory norm we're considering now would still represent a radical limitation on individual freedom of expression for the sake of limiting possible offense or trolling.

9 For the sake of simplicity, we will ignore a hybrid option proposed by Zheng et al. (2022). In such a hybrid "[u]sers can adjust both their own avatars' identities, such as gender, race, and voice, and select to hide or show specific users' avatars. For example, in VRchat, when a user encounters a user that, despite their higher Trust Rank, is wearing an avatar they feel uncomfortable, they can choose "hide avatar" in their social panel" (4). Such an option combines the issues we've associated with the first two solutions i.e., empowering users to see XR embodiment overlays while also allowing them to customize their embodiment however they wish (although this allows others to ignore these overlays, the user is empowered to see their own body how they see fit). Such a hybrid, we think, compounds the ethical issues we discuss while leaving unsolved the importance of social visibility and the problems of empowering others to erase XR embodiment/identity.

10 All corporations must operate against background laws already existing in the countries where they provide services, so, in a sense, corporate "regulation" is a subcategory of the "state regulation" approach. The major difference lies in whether states choose to set explicit standards for behavior in virtual communities that are different from those already written to regulate speech and behavior outside digital realms.

11 For example, social media communities like *Parler* have been banned by most major application storefronts due to their embracement of violent and often racist political rhetoric (Leswing 2021). Haley's (2015) play *The Nether* imagines a metaverse community in which its adult members play the role of children for their sometimes violent pedophilic clients.

12 An additional problem we'll mention but not resolve is a data privacy issue. Although intimate data related to our browsing habits and locations are real problems, the amount of intimate data that can be captured by XR technologies is significant including eye-tracking (already built into some VR head-mounted displays) can assess not only where a user is looking but also the cognitive load devoted to it by examining pupillary changes (Bozkir, Geisler, & Kasneci 2019). Heart rate, skin conductance, user height, and heart rate are all capturable in principle with present-day technology.

13 In this chapter, we wish to avoid taking sides in metaphysical disputes about the nature of the "true" self. It's enough for our purposes to show that modern concepts of the self, whatever it turns out to be, are in the grip of intense debates regarding the relationship between that self and the unaugmented or unmodified physical/natal body.

14 The ethics of trans gender confirmation surgeries, for example, has now, rightly, settled largely in the affirmative (that such surgeries are morally permissible ways of modifying the physical body to better align with a person's subjective identity) (Berli et al. 2017). While controversies remain in the case of children (Steensma & Cohen-Kettenis 2018; Sarajlic 2020), few bioethicists today argue that it is impermissible for individuals with a gender identity that differs from their natal sex to undergo confirmation surgery.

15 Some users may find this liberating, a feature not a bug. One can imagine, for example, the freedom that trying out new identities in such "burner" accounts may afford in terms of exploring new senses of gender identity, sexual orientations, etc.

16 Wouldn't it be possible for Human-like bots to fake a pulse and thus undermine our proposal? In one sense, this is obviously possible; however, we think this way of defeating our proposal is morally loaded in a way that's different from the concerns we raised over biomarkers, NFTs, or patterns of user activity.

Why? Any agent (Human or Human-like) capable of existing across metaverse spaces will need NFT key access (if those are tied to XR embodiment) and also generate patterns of engagement that uniquely identify it (as a matter of carrying out its functions). These elements would be necessary for the bot to carry out its functions. Adding a fake pulse to a Human-like bot is not an essential feature of its activity and is more akin to identity theft. It's an addition designed *only* to allow it to engage in deception. A *biological* Human-like bot (Moškon et al. 2021) would, on the other hand, force us to search for more complex generic biomarkers that differentiate Human from non-Human biological activity.

17 Even non-identifying biomarkers like these would add another layer of security and privacy risk to our metaverse existence, and it will be important that such data is safeguarded (an increased heart rate while in the metaverse, for example, could be valuable information to various corporate bodies who could target the user based on that information).

18 We can imagine a transhumanist metaverse where most agents are both biological and digital or one where agents who are born biological may come to identify far less with these biological aspects of their embodiment and far more with the digital aspects of their embodiment than they do today. Requiring generic biomarkers may thus be seen as a form of biological chauvinism in such a world.

References

American Psychiatric Association. (2022). *Diagnostic and statistical manual of mental disorders* 5th ed. Text Revision. Washington, DC.

Baron, N.S. (2013). Shall we talk? Conversing with humans and robots. *The Information Society: An International Journal*, 31 (3): 257–264.

Barras-Hill, L. (June 25, 2021). L'Oréal and Facebook use ModiFace to bring AR makeup trials to Instagram. *TRBusiness*. Retrieved from https://www.trbusiness.com/regional-news/international/loreal-and-facebook-use-modiface-to-bring-ar-makeup-trials-to-instagram/209021

Berli, J.U., Knudson, G., Fraser, L., Tangpricha, V., Ettner, R., Ettner, F.M., Safer, J.D., Graham, J., Monstrey, S., & Schechter, L. (2017). What surgeons need to know about gender confirmation surgery when providing care for transgender individuals: A review. *Journal of the American Medical Association Surgery*, April 1; 152 (4): 394–400.

Bostrom, N. (2005). Transhumanist values. *Journal of Philosophical Research*, 30: 3–14.

Bozkir, E., Geisler, D., & Kasneci, E. (2019). Person independent, privacy preserving, and real time assessment of cognitive load using eye tracking in a virtual reality setup. IEEE Conference on Virtual Reality and 3D User Interfaces (VR): 1834–1837. https://doi.org/10.1109/VR.2019.8797758.

Culver, C.M., & Gert, B. (2004). Competence. In Radden, J. (ed), *The Philosophy of Psychiatry: A Companion*. Oxford University Press: 258–271.

Cutting, J.E., & Kozlowski, L.T. (1977). Recognizing friends by their walk: Gait perception without familiarity cues. *Bulletin of the Psychonomic Society*, 9: 353–356.

Davis, J.L. (2012). Narrative construction of a ruptured self: Stories of transability on Transabled.org. *Sociological Perspectives*, 55 (2): 319–340.

de Ruiter, A. (2021). The distinct wrong of deepfakes. *Philosophy & Technology*: 1311–1332. https://doi.org/10.1007/s13347-021-00459-2.

Desatoff, S. (July 6, 2020). Pokemon Go surpasses $3.6 billion in lifetime revenue. *GameDaily.bix*. Retrieved from https://gamedaily.biz/article/1795/pokemon-go-surpasses-36-billion-in-lifetime-revenue

Dowling, M. (2021). Fertile LAND: Pricing non-fungible tokens. *Finance Research Letters*. https://doi.org/10.1016/j.frl.2021.102096

Falk, B., Meng, Y., Zhan, Y., & Zhu, H. (2021). POSTER: ReAvatar: Virtual reality de-anonymization attack through correlating movement signatures. CCS '21: Proceedings of the 2021 ACM SIGSAC Conference on Computer and Communications Security November: 2405–2407

Fischer, J.M., & Ravizza, M. 1998. *Responsibility and control: A theory of moral responsibility*. Cambridge: Cambridge University Press.

Floridi, L. (2021). Trump, Parler, and regulating the infosphere as our commons. *Philosophy & Technology*, 34: 1–5

Freeman, G., Zamanifard, S., Maloney, D., & Adkins, A. (2020). My body, my avatar: How people perceive their avatars in social virtual reality. *CHI '20 Extended Abstracts*, April 25–30, 1–8. https://doi.org/10.1145/3334480.3382923

Gillon, R. (2015). Defending the four principles approach as a good basis for good medical practice and therefore for good medical ethics. *Journal of Medical Ethics*, 41: 111–116.

Golf-Papez, M., & Veer, E. (2017). Don't feed the trolling: Rethinking how online trolling is being defined and combated. *Journal of Marketing Management*, 33 (15–16): 1336–1354.

Gros, D., Li, Y., & Yu, Z. (2021). The R-U-A-Robot dataset: Helping avoid chatbot deception by detecting user questions about human or non-human identity. Proceedings of the 59th Annual Meeting of the Association for Computational Linguistics and the 11th International Joint Conference on Natural Language Processing (Volume 1: Long Papers): 6999–7013.

Guo, C., & Caine, K. (2021). Anonymity, user engagement, quality, and trolling on Q&A Sites. Proceedings of the ACM on Human-Computer Interaction, 5 (CSCW1), Article No.: 141. https://doi.org/10.1145/3449215.

Haley, J. (2015). *The nether*. Evanston IL: Northwestern University Press.

Haslanger, S. (2000). Gender and race: (What) are they? (what) do we want them to be? *Noûs*, 34 (1): 31–55.

Higgins, S., & Wysong, A. (2018). Cosmetic surgery and body dysmorphic disorder - An update. *International Journal of Women's Dermatology*, 4 (1): 43–48.

Johnson, D.G., & Diakopoulos, N. (2021). What to do about deepfakes. *Communications of the ACM*, 64 (3): 33–35.

Kanwisher, K., McDermott, J., & Chun, M.M. (1997). The fusiform face area: A module in human extrastriate cortex specialized for face perception. *The Journal of Neuroscience*, 17 (11): 4302–4311.

Kaplan, J.T., Aziz-Zadeh, L., Uddin, L.Q., & Iacoboni, M. (2008). The self across the senses: An fMRI study of self-face and self-voice recognition. *Social Cognitive and Affective Neuroscience*, 3 (3): 218–223, https://doi.org/10.1093/scan/nsn014.

Kleemans, M., Daalmans, S., Carbaat, I., and Anschütz, D. (2018). Picture perfect: The direct effect of manipulated Instagram photos on body image in adolescent girls. *Media Psychology*, 21 (1): 93–110. https://doi.org/10.1080/15213269.20 16.1257392.

Lang, B. (February, 2021). 'Beat Saber' has sold 4M copies & 40M songs, an estimated $180M revenue (and accelerating). *Road to VR*. Retrieved from, https://www.roadtovr.com/beat-saber-4-million-units-milestone-revenue/

Leswing, K. (January 9, 2021). Apple removes parler from app store in wake of U.S. capitol riot. *CNBC*. Retrieved from: https://www.cnbc.com/2021/01/09/apple-removes-parler-from-app-store-in-wake-of-us-capitol-riot.html

Maguinness, C., & Newell, F. N. (2014). Recognizing others: Adaptive changes to person recognition throughout the lifespan. In Schwartz, B. L., Howe, M. L., Toglia, M. P. & Otgaar, H. (eds.), *What is adaptive about adaptive memory*. Oxford University Press?: 231–257.

McCabe, M., de Waal, T., & Fabri, A. (2017). Women, makeup, and authenticity: Negotiating embodiment and discourses of beauty. *Journal of Consumer Culture*, 20 (4): 656–677.

Milinski M., Croy I., Hummel T., & Boehm T. (2013). Major histocompatibility complex peptide ligands as olfactory cues in human body odour assessment. *Proceedings of the Royal Society B*, 280: 20122889. http://dx.doi.org/10.1098/rspb.2012.2889

Morton, S. (August 20, 2020). Make a more personal connection with eye contact: Now generally available. *Microsoft Devices Blog*. Retrieved from https://blogs.windows.com/devices/2020/08/20/make-a-more-personal-connection-with-eye-contact-now-generally-available/.

Moscufo, M. (March 11, 2021). Digital artwork sells for record $69 million at Christie's first NFT auction. *NBC News*. Retrieved from, https://www.nbcnews.com/business/business-news/digital-artwork-sells-record-60-million-christie-s-first-nft-n1260544

Moškon, M., Komac, R., Zimic, N., & Mraz, M. (2021). Distributed biological computation: from oscillators, logic gates and switches to a multicellular processor and neural computing applications. *Neural Computing and Applications*, 33: 8923–8938

Neely, E.L. (2019). Augmented reality, augmented ethics: Who has the right to augment a particular physical space? *Ethics and Information Technology*, 21: 11–18.

Nelson, L., & Ramirez, E. (2017). Can suicide in the elderly be rational? In McCue, R.E. & Balasubramania, M. (eds.), *Rational suicide in the elderly: Clinical, ethical and sociocultural aspects*. Springer: 1–21.

Novet, J. (June 22, 2022). Mark Zuckerberg envisions a billion people in the metaverse spending hundreds of dollars each. *CNBC*. Retrieved from: https://www.cnbc.com/2022/06/22/mark-zuckerberg-envisions-1-billion-people-in-the-metaverse.html

O'Leary, D. (2019). GOOGLE'S Duplex: Pretending to be human. *Intelligent Systems in Accounting, Finance, and Management: An International Journal*, 26 (1): 46–53.

Paul, L.A. (2016). *Transformative experience*. Oxford: Oxford UP.

Pradhan, A., & Lazar, A. (2021). Hey Google, do you have a personality? Designing Personality and Personas for Conversational Agents. CUI '21: Proceedings of the 3rd Conference on Conversational User Interfaces, Article No.: 12, 1–4. https://doi.org/10.1145/3469595.3469607

Ramphul K., & Mejias S.G. (March 03, 2018). Is "Snapchat Dysmorphia" a real issue? *Cureus* 10 (3): e2263. https://doi.org/10.7759/cureus.2263.

Rice, S.M., Graber, E., & Kourosh, A.S. (2020). A pandemic of dysmorphia: "Zooming" into the perception of our appearance. *Facial Plastic Surgery & Aesthetic Medicine*, 22 (6): 401–402.

Sarajlic, E. (2020). Children, culture, and body modification. *Kennedy Institute of Ethics Journal*, 30 (2): 167–190. https://doi.org/10.1353/ken.2020.0005.

Schmeiss, J., Hoelzle, K., & Tech, R.P.G. (2019). Designing governance mechanisms in platform ecosystems: Addressing the paradox of openness through blockchain technology. *California Management Review*, 62 (1): 121–143.

Steensma, T.D., & Cohen-Kettenis, P.T. (2018). A critical commentary on "A critical commentary on follow-up studies and "desistence" theories about transgender and gender non-conforming children." *International Journal of Transgenderism*, 19 (2): L 225–230.

Stone, R.J. (2000). Haptic feedback: A brief history from telepresence to virtual reality. In Brewster, S. & Murray-Smith, R. (eds.), *Haptic human-computer interaction*. Springer: 1–16.

Su, J., Shukla, A., Goel, S., Narayanan, A. (2017). De-anonymizing web browsing data with social networks. WWW '17: Proceedings of the 26th International Conference on World Wide Web: 1261–1269.

Sugiura, M., Watanabe, J., Maeda, Y., Matsue, Y., Fukuda, H., & Kawashima, R. (2005). Cortical mechanisms of visual self-recognition. *NeuroImage*, 24 (1): 143–149.

Tremblay, S.C., Tremblay, S.E., & Poirier, P. (2021). From filters to fillers: an active inference approach to body image distortion in the selfie era. *AI & Society*, 36: 33–48.

Tuvel, R. (2017). In defense of transracialism. *Hypatia*, 32 (2): 263–278.

Vi, S., da Silva, T.S., & Maurer, F. (2019). User experience guidelines for designing HMD extended reality applications. In: Lamas, D., Loizides, F., Nacke, L., Petrie, H., Winckler, M. & Zaphiris, P. (eds.), Human-Computer Interaction – INTERACT 2019. INTERACT 2019. Lecture Notes in Computer Science, vol 11749. Springer.

Vita-More, N. (2021). The Body vehicle: An argument for Transhuman bodies. In Gray, C.H., Figueroa-Sarriera, H.J. & Mentor, S. (eds.), *Modified: Living as a cyborg*. Routledge: 58–67.

Weber-Guskar, E. (2021). How to feel about emotionalized artificial intelligence? When robot pets, holograms, and chatbots become affective partners. *Ethics and Information Technology*, 23: 601–610.

Westerlund, M. (2019). The emergence of deepfake technology: A review. Technology Innovation *Management Review*, 9 (11): 40–53. http://doi.org/10.22215/timreview/1282

Zheng, Q., Wang, L., Ngoc Do, T., & Huang, Y. (2022). Facing the illusion and reality of safety in social VR. CHI EA '22, Proceedings of the 1st Workshop on Novel Challenges of Safety, Security and Privacy in Extended Reality, April 29–May 5. New Orleans, LA, USA.

7 Moral Narratives in Virtual Worlds

Andrew Kissel

Discussions of morality in connection with video games have historically focused on the extent to which violence in video games leads to violence outside of video games. As of February 2020, the American Psychological Association (APA) acknowledges an association between violent video game play and heightened aggression.[1] However, the APA resolution stresses that there is no known connection between violent video game play and real-world criminal or lethal violence. Indeed, the idea that enjoying violent video games will make a person more violent has been generally dismissed.

Recent work by philosophers has pursued a slightly different question. They ask, "Can playing a video game be morally problematic in itself, independent of further real-world outcomes?"[2] Intuitive responses to this question are mixed. On the one hand, there appears to be a widespread tendency to shrug off otherwise immoral behavior in virtual contexts because "It's just a game." Video game actions are *fictional*, so the thinking goes. On this view, so long as no one is *actually* harmed, video game violence is an acceptable form of imaginative fun. Gaming culture largely reflects this intuition. Games built around battles to the death, such as *Fortnite* and *Apex Legends*, are among the most profitable and popular contemporary video games.

On the other hand, games that include sexual deviance and sexual violence remain on the fringes of game culture. Upon its announcement in 2019, the controversial single-player computer game *Rape Day* promised to allow players to "verbally harass, kill people, and rape women as you choose to progress the story."[3] Many in the gaming community found the idea of enjoying that kind of content deeply repugnant, despite the developer arguing that "You can't reasonable [sic] consider banning rape in fiction without banning murder and torture...".[4] The game was ultimately pulled from the Steam online store.

The reaction to *Rape Day* suggests that, at least sometimes, players make moral evaluations on the basis of actions that are limited to virtual

DOI: 10.4324/9781003359494-10

contexts.[5] Philosophers that advocate *willing endorsement* approaches to the ethics of video games hold that such moral evaluations are justified.[6] For example, Christopher Bartel argues that players are candidates for moral blame when they perform in-game actions that they *endorse* via their real-world moral psychology.[7] In a similar vein, Stephanie Patridge points out that we would radically alter our view of a close friend when we discover that they regularly enjoy engaging in VR games that simulate pedophilia, even if we knew that they did not engage in pedophilia in real life. Their enjoyment of such games reveals a "failure both of sensitivity and sympathy," which she calls "an obvious vice of character".[8] These philosophers are united in their view that virtual actions can be indicative of the *character* of a player when the player willingly endorses those actions through their behavior. In which case, players are appropriate targets of moral evaluation on the basis of their virtual actions, independent of further consequences.[9]

Willing endorsement approaches to the ethics of video games hinge crucially on an explanation of the connection between virtual actions and a player's character. Presumably, not all virtual actions reflect the genuine characteristics of the player. Players often enjoy video games precisely because they can act in *uncharacteristic* ways, free from the strictures of reality. This observation has led willing endorsement approaches to distinguish between those actions that reflect a player's genuine moral characteristics (in some sense) and those actions that merely reflect fictional roles that the player adopts for gameplay purposes.[10] The thinking goes that only actions that reflect a player's genuine characteristics can support moral evaluations.

It is against this background that the rising popularity of virtual reality (VR) technologies is particularly intriguing from an ethical perspective. VR technologies increase the possibility of what philosophers Erick Ramirez and Scott LaBarge call *virtually real experiences*.[11] "Such experiences are treated by subjects as if they were real experiences (via some combination of behavioral, physiological, neurological, or cognitive similarities between virtual and real experiences)."[12] If a VR experience is virtually real, then user reactions to that experience plausibly reflect their genuine characteristics. As such, the increased probability of virtually real experiences presented by VR also increases the probability that virtual actions (and the users that perform them) are appropriate targets of moral evaluation.

In this paper, I argue that determining when a virtual action reflects genuine characteristics of the user requires an answer to the *Virtual Characterization Problem* (VCP), the problem of determining the conditions under which a virtual action is properly attributable to the user. This is a problem of personal identity. I maintain that virtual actions ground moral evaluations only in those cases where the action is properly attributable to the user.

I then argue that the binary distinction between "genuine characteristics" of users and virtual "roles" of users adopted by willing endorsement views fails. It fails, in part, because it does not acknowledge the way that virtual roles can *themselves* constitute a user's identity.

I draw on narrative views of identity to sketch an alternative answer to the VCP. On this view, virtual actions ground moral evaluations to the extent that those actions are properly attributable to this complex, role-based (and partly virtual) identity.

The upshot of this view is that the ethical impact of VR technologies, in terms of grounding moral evaluations, is largely *independent* of their ability to generate virtually real experiences. Virtual roles may be constitutive of one's identity even when the software does not generate a virtually real experience. Nevertheless, since VR technologies provide greater opportunities for users to exercise agency, they afford greater opportunities to constitute narrative identity. In which case, they still offer greater opportunities for moral evaluation, as predicted by willing endorsement views. The resulting picture is a character-based approach to the ethics of video games that can also address subtleties of identity in virtual contexts.

The Virtual Characterization Problem

In his 2020 book *Video Games, Violence, and the Ethics of Fantasy*, Christopher Bartel provides the following summary argument for when it is appropriate to judge a player for their virtual actions:

1 Some desires are immoral to possess even when they are not acted upon.
2 Desires can be cultivated, either in reality or through fantasy.
3 It is morally wrong to cultivate an immoral desire.
4 Video games can be used as a prop for one's fantasies.
5 As such, they can be used to cultivate desires, both moral and immoral.
6 Therefore, it is morally wrong to play a video game when doing so serves to cultivate an immoral desire.[13]

Crucially, for Bartel, immoral desires must be components of our *genuine* moral psychologies. They must be desires that could be effective in *real-world* contexts, even if we never act on them in those contexts. And importantly, these desires, attitudes, and moral psychologies are partly constitutive of our sense of self.[14] Bartel writes, "It is not 'just a game.' It is a fantasy about how some people want to see the real world...Our fantasies are part of our character, and our pretend play offers a way to engage and reinforce our fantasies."[15]

Bartel's argument provides a template for the more general class of views that I call *willing endorsement* views. Willing endorsement views hold that it may be appropriate to evaluate a player on the basis of their virtual actions only when those actions reflect the player's *genuine moral characteristics*, where "genuine moral characteristics" here seems to mean those character traits that are brought to bear in real-world (i.e., non-virtual) contexts.

For willing endorsement views, then, morality in virtual contexts depends crucially on the player's relationship to their actions. When players act out of genuine immoral desires, they are morally culpable. For example, players of *Rape Day* are generally seen as morally repugnant on willing endorsement views because the players who enjoy *Rape Day* seem to be acting on *genuine* desires to sexually assault someone, even if they have never (and would never) sexually assault someone in real life. By contrast, players who enjoy games like *Fortnite* and *Apex Legends* are NOT acting on genuine desires to murder, the story goes, and so can be viewed as enjoying harmless fun.

An interesting upshot of the willing endorsement view is that the same action could ground completely different moral evaluations for two people who nevertheless perform the same virtual action. Consider the following scenario

Taika & Fatima

Taika and Fatima are playing the Bethesda Softworks game *The Elder Scrolls V: Skyrim VR* (2018) together. They take turns engaging in various fantasy tropes as the main protagonist. At Taika's suggestion, Fatima uses her high "Sneak" ability to quietly slit the throat of a non-player character (NPC) in the town of Winterhold. Both players had found the NPC annoying. Fatima chuckles when no one in the town reacts (due to the high "Sneak" level). Taika, however, laughs uncontrollably and excitedly takes the controller to try it himself. "I've been wanting to try this for a while!" he exclaims as he begins to assassinate as many NPCs in the town as he can without getting caught. Taika's enthusiasm makes Fatima uncomfortable and she begins to think that he might not be the kind of person she thought he was.

According to willing endorsement views, Fatima's reassessment of Taika's character may be justified. Taika's virtual actions plausibly reflect something genuine about his moral character, perhaps that he is the kind of person who enjoys viewing the sadistic murder of innocents. Although he might never murder someone in real life, his murder of the virtual NPCs helps satisfy a genuine desire for murder had by Taika. By contrast, Fatima recognizes that she is just pretending to be an assassin. Plausibly, her virtual actions do NOT reflect something genuine about her moral character. In which case, the willing endorsement view holds that it is appropriate to evaluate Taika on the basis of his virtual actions but not Fatima.[16]

At this point, one might argue that moral criticism of Taika on the basis of his virtual actions is always unjustified by rejecting the virtue ethics underpinnings of willing endorsement views. Since no one is harmed by Taika's actions, they may say he has done nothing wrong. To provide a full defense of virtue ethics would go far beyond the scope of this paper.[17] For present purposes, one need only accept that it is sometimes appropriate to evaluate a person on the basis of their moral character, independent of the further negative consequences of their actions.

What is in need of further defense, however, is the claim that some (though not necessarily all) virtual actions reflect the *genuine* moral characteristics of a player. Plausibly, most virtual actions *do not* reflect the genuine characteristics of players. Players like Fatima *pretend* to be someone different in virtual contexts from who they are in reality. So what makes an immoral desire a genuine characteristic of a player rather than a bit of pretend adopted for harmless fun? We can call this the *virtual characterization problem*.

> **Virtual Characterization Problem (VCP):** What are the conditions under which virtual actions (and the desires that drive them) are *properly attributable* to players?[18]

What does it mean for an action to be "properly attributable" to a player? As a first pass, an action is properly attributable to a person when it reflects the genuine characteristics of the person *that make her who she is*. The concept comes in various guises in philosophical literature. Harry Frankfurt, for example, argues that an action is properly attributable to a person when the principle behind the action is located within that person and is one with which they identify.[19] Gary Watson initially argued that proper attribution requires evaluation of the action on the part of the agent in terms of their conception of the good.[20] He later believed that actions were properly attributable to an agent only when they involved brute self-assertion on the part of the agent.[21] Charles Taylor and others hold that an action is properly attributable when it is correctly included in the narrative the agent talks about themselves.[22]

All of the above views agree that an action may be *causally* attributable to a person without necessarily reflecting the genuine moral characteristics of that person. For example, a son may excuse the hurtful words of his father with dementia by saying, "That's not him speaking, that's the disease speaking." In doing so, the son acknowledges that his father is the causal source of the hurtful words while denying that they are *properly* attributable to his father.[23]

It is important to note that proper attribution admits of degrees. It may be the case that the father's hurtful words are wholly out of character due to his dementia, in which case the son's response seems appropriate. However, it could instead be the case that the father has a long history of saying hurtful things about his son. While the dementia may lead the father to speak out when he might have otherwise held his tongue, there is still a sense in which the hurtful words reflect some aspect of the father's genuine character. In which case, the hurtful words are more properly attributed to the father than if he did not have that history, but not as properly attributable as when he speaks without any influence from dementia at all.

In virtual contexts, it is rarely difficult to determine the causal source of an action. The player who selects the inputs is the causal source. It is far harder to determine to what extent an action is properly attributable to the player. The issue I wish to focus on here is not the epistemic one of how we could *know* whether the action is properly attributable. Instead, I will focus on the issue of *what makes it the case* that a virtual action is more or less properly attributable to a player.

Addressing the VCP is thus crucial for the ultimate success of willing endorsement views. Without some criteria to determine what makes an action properly attributable to a player, it is hard to see how players like Fatima are morally different from players like Taika. After all, both act on a desire to kill innocents in a virtual context. Furthermore, both are fully aware that they are *not actually* assassins. So, what makes Fatima's pretending different from Taika's pretending?

An initially attractive option is to consider the possibility that Taika acts on real-world desires. That is, perhaps Taika acts on a desire aimed at seeing innocents killed in reality, not just virtually, even if he has never acted on that desire in real life. In the next section, I argue that the simple distinction between characteristics of a player that reflect real-world desires and those that reflect mere pretend is not robust enough to answer the VCP.

The Simple Roleplaying Response

Defining Simple Roleplaying

In non-video game contexts, determining *whether* an action is properly attributable to a person tends to be fairly straightforward. Dementia, addiction, coercion, etc., are exception cases where we do *not* think an action is properly attributable to the immediate cause of the action. But in most cases, the person who is the cause of action is also the person to whom the action is properly attributable.

In virtual contexts, however, it seems much more difficult to determine which actions are properly attributable to the player. The difficulty is that virtual worlds often allow players to engage in *simple roleplaying*. A player engages in simple roleplaying when they perform a virtual action *because* that is what they think is dictated by the fictional role they have adopted in the game.[24] For example, a Skyrim player engaging in simple roleplaying might choose to resort to violence to complete a mission because that is what the player thinks an assassin, acting in character, would do, even if the player also thinks that the violence would be morally wrong, where the situation to occur in a non-virtual context. Simple roleplaying contrasts with players who act in order to complete the mission as efficiently as possible, or to achieve a particular ending, etc.

Simple roleplaying is not limited to virtual contexts. Some actors might use simple roleplaying as a technique for portraying their characters. In the TV show *Extras*, the actor Ian McKellen describes his process of portraying the wizard Gandalf as follows: "I imagined what it would be like to be a wizard, and then I pretended and acted in that way on the day."[25] Although the description is tongue-in-cheek, it describes simple roleplaying. Very few players *really* think they are assassins, cowboys, orcs, or space aliens. Instead, players generally *pretend* that they are these various characters through simple roleplaying. Of course, it may be the case that what is dictated by the fictional role *coincides* with what the player would do were the situation actual. However, that does not mean the player is performing the action *because* it is what they would do were the situation actual.

Simple roleplaying is not always a stable state maintained over an extended virtual session. Players may choose to perform an action because it fits with the character of their in-game avatar... and then a moment later perform an action *contrary* to that character because it will be the most effective way to complete the game successfully. The latter is not simple roleplaying because the action is dictated by considerations outside of those dictated by the fictional role adopted by the player.

It is important for simple roleplaying that the adopted role be dictated by the internal fiction of the game. Roles dictated by the fiction of the game include the stories, environments, characters, etc., created by the game's development and design teams. It also includes the fictional elements contributed by players through their play, such as when a player develops their own character in a massively multiplayer online game. However, not all roles a player adopts in the course of a play session are dictated by the fictional elements of the game. A player might decide to play a round of a game as a "griefer." Griefers intentionally harass and provoke other players in order to decrease their enjoyment of the game. The griefer role this player adopts is a meta-game role, not a role dictated by the fiction. As such, griefing does not involve simple roleplaying, though it does involve adopting a role.[26]

Simple Roleplaying and the VCP

Players engage in simple roleplaying when they act on the basis of what is dictated by the fictional role they adopt rather than on the basis of what they themselves would do were the situation real. Since actions that result from role-based pretend play need not reflect genuine moral characteristics on this view, virtual actions are properly attributable (in the sense discussed earlier) only to the extent that the player does *not* engage in simple roleplaying, *ceteris paribus.*[27]

Let's see how well simple roleplaying can explain the difference in moral judgments between Fatima and Taika. Plausibly, Fatima engages in simple roleplaying. She first considers which behaviors are dictated by the role of "assassin." She then performs virtual actions that match what an outlaw assassin might do in real life *on that basis.* Fatima kills the NPC train tracks *because* she is adopting a fictional role. While she enjoys roleplaying, her virtual actions do not reflect genuine characteristics that are properly attributable to her (aside from the fact that she enjoys roleplaying). Rather, her actions are attributable to the fictional role that she has adopted. Crucially, we can see that Fatima's actions are a form of simple roleplaying because, ex hypothesi, she has no desire to see real-life innocents killed. Actions and desires involved in simple roleplaying aim at strictly virtual contexts, while genuine moral characteristics aim at non-virtual (i.e., real-world) contexts.

Taika, the story continues, is an appropriate target of moral evaluation, at least to the extent that the virtual murder of the farmer reflects genuine characteristics that are properly attributable to him. Taika murders the NPCs because he enjoys hearing them wail, watching the virtual blood spatter across the ground, etc. Plausibly, Taika enjoys these experiences because they tap into a genuine, real-world desire to see innocents killed. Of course, Taika might never act on this real-world desire because he recognizes that the killing of innocents is wrong. But the killing of virtual innocents provides the best proxy to satisfy his real-world desires. In short, Taika enjoys the sadistic murder of innocents, independent of any role he has adopted in the game. The game merely affords him a means to satisfy this desire. In this case, since Taika acts on a real-world-directed desire rather than a pretend desire acquired through simple roleplaying, his virtual actions reflect his genuine moral character.

In summary, it is intuitively appropriate to morally evaluate Taika, but not Fatima, on the basis of his virtual actions. The simple roleplaying response explains this difference. Fatima roleplays, while Taika does not, and so the action is only properly attributable (for the purposes of moral evaluation) to Taika. In which case only Taika should be morally evaluated for his virtual actions.

If the simple roleplaying response is correct, then the increasing popularity of VR and VR games should be accompanied by an increase in opportunities for the moral evaluation of players on the basis of their virtual actions. Successful VR environments are highly *immersive* and engender a strong sense of *presence* in the user. In the VR literature, *presence* refers to "the sense of residing in the simulated environment with an illusion of being transported to a place/time different than where the user actually is."[28] Immersion refers to the aspects of the experience itself that can (though need not) contribute to the experience of presence. The immersive qualities of VR and their ability to create increased experiences of presence contribute to the possibility of *virtually real experiences*.[29] VR creates experiences for user's experiences that they treat *as if they were real*. When a VR user stands at the edge of a virtual cliff, their hands may start to sweat and their knees may start to shake. Of course, if you ask the user whether they are actually standing on a cliff, they will answer in the negative. Nevertheless, they continue to treat the experience as if it were real.

Evidence from psychology suggests that VR is very good at creating virtually real experiences, to the point that it has been used to run experiments that otherwise would be impossible, such as a VR replication of the famous Milgram shock experiment.[30] Similarly, Francis et al. use VR to study moral behavior, as opposed to moral judgment, in response to life and death decision dilemmas.[31] The lesson seems to be that if we can create a virtually real experience in VR, then we can learn something about a person's *actual* moral behavior. On the simple roleplaying view, the closer a VR experience comes to generating a virtually real experience, the more likely it is that the player is acting on their genuine moral characteristics, and therefore the more appropriate it is to make moral evaluations of the player. Or so the story goes.

Problems for the Simple Roleplaying Response

Despite its initial intuitive appeal, the simple roleplaying response to the VCP is *too* simple.[32] It assumes that desires aimed at the real world are genuine characteristics of a person's identity. Meanwhile, desires resulting from an adopted role are viewed as mere "pretend." But such a view fails to acknowledge the complexity of contemporary identities. More specifically, the simple roleplaying response fails to acknowledge that adopted roles are often *constitutive* of one's identity. In those cases where an adopted role is constitutive of one's identity, an action resulting from an adopted role may be properly attributable to an individual, including roles adopted in strictly virtual contexts.

In order to explain the difference between Taika and Fatima, the simple roleplaying response divides the videogame player's identity into two. On the one side is the "genuine" or "true" identity that is the player's default identity deployed in reality. On the other side is the "pretend" or "role-played" identity that is deployed in the context of playing Skyrim VR. Actions rooted in the player's true identity ground moral evaluations because they are properly attributable to the player, while actions rooted in the roleplayed identity do not.[33] The result is that actions resulting from such roleplayed identities are *never* properly attributable to the player.

In some cases, however, we seem to view ourselves as playing a role while also viewing that role and the actions that flow from that role as properly attributable to ourselves. This point is particularly noticeable when the roles one adopts conflict. Consider the case of Chris, a person who identifies themselves both as a Catholic and as a member of the LGBTQIA+ community.[34] In Catholic settings, Chris plays up the Catholic aspects of their identity while suppressing their LGBTQIA+ identity. For example, Chris might admit that they find something shameful about certain kinds of sexual practices. Such an admission is Chris leaning into their Catholic identity role. In contrast, Chris might express an interest in sexual exploration and experimentation among LGBTQIA+ friends. In this case, Chris is giving voice to their LGBTQIA+ identity. Importantly, both roles (and the accompanying social behaviors) are core aspects of Chris' identity. Chris is not being two-faced. Chris is not secretly endorsing one role while only pretending to endorse another. Rather, both roles contribute crucially to Chris' understanding of their own identity, an identity that includes a painful conflict that must be navigated.[35]

In the case of Chris, it would not make sense to say that one of these roles is the "true" identity while the other is a "pretend" identity. Both roles contribute to their self-identity in important ways. So, the mere fact that a person has adopted a role that contributes to their behavior does not, in itself, indicate that the action is not properly attributable to the person. By extension, the mere fact that a player adopts one role in virtual contexts and another role in other contexts does not, on its own, give any indication that one role is *more* genuine than the other.

Still, it may be argued that the roles adopted in virtual contexts are importantly different from those adopted in the real world. Specifically, the roles that video game players adopt are contained within *fictional* worlds.[36] It is not the fact that Fatima adopts a role that makes her free from moral criticism. Rather, it is the fact that she adopts a *pretend* role within a fictional world. Roles adopted in fictional contexts just aren't the sort of things that are properly attributable to individuals.

I find the above argument puzzling. Many children's games and educational games are predicated on the idea that children can develop good moral character *by way* of pretend play rather than *despite* it. There is extensive evidence that pretend play, and playing games more generally, is crucial for healthy childhood development.[37] Roleplaying has long been used successfully in therapy and counseling.[38] Since fictional roles can aid in moral development, it seems reasonable to me to think that fictional roles can be properly attributable to a person.

As a final criticism of simple roleplaying, it is important to recognize that it is possible to roleplay as a character, in the simple sense that you perform the action *because* it is what the character would do, while also thinking that the action is the correct one, all things considered. In such cases, what the player does, qua roleplayer is the same as what the player would do were they *not* roleplaying. In this case the roleplayed action seems to reflect the genuine characteristics of the player themselves.

A defender of simple roleplaying might point out that although a role-played action may be identical to the action the player would perform (were they not roleplaying), the roleplayed action is not related to the gen-uine characteristics of the player in the right way. That is, we might think that it matters whether you performed the action *because* it was dictated by a role rather than being caused by genuine moral characteristics. In this sense, the defender might point out simple roleplayed actions are not caused by the player's *true* identity (in the right way). This means that as long as a player engages in simple roleplaying, it is inappropriate to revise our moral evaluations of a player on the basis of a virtual action, even when that action matches what they would do were the scenario actual.

The defense in the previous paragraph gets to the heart of the difficulty with the simple roleplaying response to the VCP. It *assumes* a divide be-tween the player's genuine moral characteristics and the roleplayed iden-tity in advance in order to explain when a player is an appropriate target of moral evaluation. That assumption requires that we already know what it is that separates the player's true identity from any roleplay identity they might adopt. But that is precisely the question the VCP is asking: in virtue of what is it appropriate to say that some collection of virtual actions is properly attributable to the player *in the first place*? Simple roleplaying does not provide an answer to the VCP. It assumes that we already know the answer.

In summary, the simple roleplaying response fails to see the complex ways that adopted roles play out in everyday life. The binary distinction between the true self and the roleplayed self is *too* simple. Identities are not always single, unified lists of "genuine" traits. Rather, identities can be dy-namic collections of intersecting and conflicting roles. The conflict in these roles may be *constitutive* of a person's identity, as seen in the case of Chris. Although these reflections are true of real-world identities and actions, they apply just as well in virtual contexts. Fictional roles can play a crucial part in moral development, justifying praise as much as blame. If we wish to know which in-game actions ground moral evaluations, we should not ask, "Is the action attributable to the player's true identity or merely their pretend, roleplayed identity?", as suggested by the simple roleplaying response. Instead, we should ask, "Is the role the player is currently adopting constitutive of their identity?" In the next section, I offer an initial sketch of how to answer this question in virtual contexts.

Narrative Identity and Robust Roleplaying

Let's take stock. The paper began by identifying the problem of determining when moral evaluations of a player are appropriate on the basis of in-game actions. Intuitively, moral evaluations are appropriate when the action reflects the genuine moral characteristics of the player. This in turn gave rise to the VCP: what are the conditions under which virtual actions (and the desires that drive them) are *properly attributable* to players? I then argued that simple roleplaying fails to fully address the VCP because it relies on an overly simplistic, binary view of self-identity.

Instead of thinking about self-identity in terms of "true" versus "pretend" selves, we should think in terms of the numerous roles we adopt and the extent to which those roles are constitutive of identity. The sort of view I am advocating for here is a species of the *narrative view* of identity. Narrative views hold that we constitute our own identities by understanding ourselves in terms of an autobiographical narrative. By understanding ourselves in terms of the roles that we adopt, we construct a narrative that links together the disparate events, actions, intentions, etc., in our lives. Self-identity is not reducible to any one of these roles but rather emerges out of the interweaving of those roles into a unified, coherent narrative.

Before expanding further on the kind of narrative view I am advocating for here, it is important to note that there are numerous narrative views of identity on offer in the literature.[39] Each has its respective strengths and weaknesses, and to defend any one version in great detail here would go far beyond the scope of the paper. For present purposes, I only seek to provide a thumbnail sketch of the broad kind of view I have in mind in order to show how it addresses the VCP more effectively than simple roleplaying views.

On narrative views, self-identity is individuated not by psychological continuity but by continuity in the form of a narrative.[40] The idea of the narrative is quite literal, though it should not be tied too closely to particular narratives from literary history. One conceives of their life in terms of a narrative when they see it as having the logic and form of a story– the story of their life.[41] Of course, one need not see themselves in terms of the "Hero with 1000 Faces", or some explicit narrative in order to constitute themselves.[42] For that matter, they need not view themselves in explicitly *narratological* terms at all. But there must be some semblance of a story, characterized by a linear sequence of events, that ties who they are now to who they have been in the past and to who they are becoming in the future. In doing so, the narrative provides a contextual frame for understanding and interpreting the events of one's life. As such, the limits and unity of the narrative determine the bounds of an individual person's identity.

We can illustrate how narrative roles construct identity by returning to the case of Chris. Chris understands and makes sense of their own experiences, past and present, in terms of the roles of *Catholic* and *LGBTQIA+ members*. In viewing themselves in these roles, they begin to create a narrative that explains their behavior. They go to church on Sunday because they are Catholic. They march in pride parades because they are LGBTQIA+ members. By adopting these roles, Chris places themselves in a web of social institutions and interactions that gives shape to their life. They construct a self-told narrative through their roles, and so they construct their own identity.

Action is properly attributable to a person when it can be made sense of in terms of the ongoing narrative that person constructs for their life. Chris' church attendance *makes sense*, to them and to others, by virtue of their Catholic narrative. As such, the action of attending church is properly attributable to Chris. In video game contexts, then, we must ask whether it is ever possible to incorporate virtual actions into the player's ongoing self-narrative. If so, then that action is properly attributable to the player.[43]

My answer is yes, so long as the player engages in *robust roleplaying*. A player engages in robust roleplaying when they adopt the in-game character's narrative as their own narrative. That is, the player not only performs the action because it is what they think the character *would* do but also they take the character's in-game narrative to be constitutive of their own self-identity. For example, social virtual reality experiences such as Meta's Horizon Worlds or AltSpace VR allow users to interact with each other in social settings that mirror non-virtual interactions. In some cases, users adopt personas, along with corresponding avatars, that do not necessarily reflect their behavior and looks in non-virtual contexts. Nevertheless, adopting these personas and avatars allows users to explore aspects of their own identities. In one study, a user with a physical impairment commented, "I'm usually stuck at home on the bed. So social VR has opened up a whole new world. I like that I get to go to real places."[44] Another user comments, "The best part of social VR I've seen so far is that I can practice my social skills… I feel more confident just by saying things I normally wouldn't. And it's a benefit."[45] The avatars and personas that users create in these social experiences allow them to explore aspects of themselves that they have difficulty manifesting in non-virtual contexts. As such, the narratives they tell about themselves in these virtual contexts should be understood as part of the broader narrative that constitutes their identity, even if they do not (or cannot) manifest them in non-virtual contexts.

We should be careful, however, not to exaggerate the adoption of a character's narrative. A player can engage in robust roleplaying without being delusional and thinking they really *are* the character they control. There are real-world constraints on the kinds of narratives a person can talk about themselves. As much as I might want to convince myself that I am a flying superhero, the basic laws of nature prevent me from attaining that goal. Similarly, no amount of Taika telling himself that he is an assassin will make it the case that he is an assassin *outside* of Skyrim VR. Nevertheless, Taika can take his character's goals and story arc as his own in virtual settings. The key is that when Taika robustly roleplays, what he does in the virtual world is part of the larger narrative that he talks about his life.

There is some evidence that people see the roles they adopt in virtual settings as being constitutive of their identities. Professional players of the videogame *Overwatch 2* might refer to themselves as "Tracer-main" to indicate that in virtual contexts, they play as the character Tracer.

Robust Roleplaying and the VCP

With an understanding of narrative identity and robust roleplaying from the previous section in hand, let us return to the question raised by the VCP. The question is: which virtual actions are *properly* attributable to a user? Narrative theories of identity and robust roleplaying provide a simple solution: those actions that are part of one's own complex, narrative-based identity is properly attributable. And crucially, one's identity is not limited to the collection of beliefs, desires, abilities, etc., that one deploys in non-virtual contexts. Virtual identities are partly constitutive of one's narrative as well, provided one is robustly roleplaying.

Robust roleplaying explains the difference between our intuitive assessments of Fatima and Taika. Although Fatima is pretending to be an assassin in Skyrim VR, that role is not part of her narrative identity. Taika, on the other hand, takes on the assassin role as part of his narrative identity. As such, Taika is an appropriate target of moral evaluation, while Fatima is not. The fact that Taika's actions are constitutive of him as a person can be seen by the fact that he continually returns to the same action. Killing the NPC is not a once-off experiment or a bit of fun in a game. Rather, it is an action that helps us understand the narrative of Taika's life. He is the kind of person who takes pleasure in the suffering of innocents, even if only in virtual contexts. When Taika kills the farmer, he is engaging in robust roleplaying. It is because of *this* fact about his life narrative that the virtual killing is attributable to him, and he is an appropriate target of moral evaluation.

It is one thing to say that Taika is an appropriate target for moral evaluation. It is quite another to say that the appropriate evaluation of Taika is a *negative* one. After all, assuming Taika's behavior is restricted to virtual contexts only, then no one is being harmed by his behavior. So, while he may be an appropriate target of moral evaluation, we may say that the appropriate evaluation is neutral. In which case, Taika is neither praiseworthy nor blameworthy for the virtual actions that are properly attributable to him.

To fully respond to this concern would require a full-blown defense of non-consequentialist ethics in virtual contexts. However, it is worth remembering two points. First, as illustrated earlier, defenders of simple roleplaying also adopt a broad virtue ethics approach. What matters for present purposes is not necessarily whether further harm is caused, but rather what the behavior plausibly reveals about the character of the user. Second, even if virtual roles have little crossover into other domains of life, I do not see why this means that in-game narratives are not constitutive of a person's identity. We do not think that a person who endeavors to keep their public and private lives separate must deny that one of those aspects of their lives is constitutive of their identity. So why should we deny that their virtual personas are partially constitutive of their identities? Adopting the narrative of an in-game character does not require that you fail to see your actions as taking place within the context of a game. And although having real-world consequences outside of the game may be good evidence that the game narrative is partially constitutive of the player's identity, it is not a necessary requirement.

The hours and effort that many users of *Skyrim VR* put in can help illustrate the substantial role virtual worlds can play in a person's life. Imagine a grown adult male married with a steady job. Every night when he gets home, instead of watching television, he plays a few hours of Skyrim, exploring a completely virtual world full of NPCs. He does not interact with any *real* people since the game is not online. Given the hours the man has put into the game, he has developed a complex narrative for his virtual avatar. Each night, he extends the character's narrative. Despite his devotion to Skyrim, it does not have a substantial impact on his work or family life. He does not discuss Skyrim with friends or family, and he does not go to Skyrim community meetups or conventions. He is like a reclusive woodworker: every night he goes down into his shop to create but leaves his work in his shop. When this man thinks about the narrative he constructs for his own life, it would be impossible to leave out the fact that he plays Skyrim. It is part of who he is as a person. Not just the fact that he *plays* Skyrim, but the particulars of the intricate narrative he has developed for his character in the game as well. Telling this man's story would be incomplete without his exploits in Skyrim. His identity would be incomplete without Skyrim.

Conclusion

This paper has endeavored to begin answering what I have called the VCP, the problem of determining the conditions under which virtual actions are properly attributable to users in virtual contexts. Answering this question is crucial for determining when moral evaluations are appropriate on the basis of virtual actions. I have endeavored to take a first step toward addressing this problem by developing the idea of robust roleplaying. When a player adopts their virtual character's narrative as their own, that narrative partially constitutes their identity. In which case the player is an appropriate target for moral evaluation. There are many questions that are still in need of answering if this sketch is to be developed more fully. Some of these questions are questions for narrative views of identity more generally. For example, what counts as a narrative? Other questions arise in the specific attempt to apply the theory to virtual contexts. For example, what is the nature of the relationship between the in-game character's identity and the player-as-character identity?

Still, I think this preliminary sketch helps illuminate the path forward while also closing down some dead ends. Attempts to draw a strong distinction between "true" identity and "pretend" identity start off on the wrong foot. And one can adopt a virtual narrative without leading to serious consequences outside the context of the game. I hope these points will provide helpful constraints in further developing the concept of robust roleplaying in the future.

Notes

1 APA (2020) https://www.apa.org/about/policy/resolution-violent-video-games.pdf
2 Patridge (2011), Bartel (2012, 2015, 2020), Luck (2009, 2018); Ali (2015, 2022); Young (2017); Heinrich (2021); Goerger (2017).
3 Hernandez (2019).
4 Ibid.
5 Luck (2009).
6 Kissel (2020).
7 Bartel (2015).
8 Patridge (2011, 310).
9 I sometimes use the term "player" to refer to the user in virtual contexts. However, the arguments presented are meant to apply to all virtual experiences, including those that may not be considered "play" or "game".
10 Bartel (2015); Patridge (2011), Ostritsch (2017), Ali (2015).
11 Ramirez and LaBarge (2018).
12 Ramirez and LaBarge (2018, 252).
13 Bartel (2020, 100).
14 Bartel (2020, 88).
15 Bartel (2020, 119).

16 Of course, Fatima may be deceiving herself about her enjoyment of the murder. And she may be mistaken about what Taika's actions reveal about his character. But this kind of epistemic uncertainty occurs in most cases where moral evaluations are under consideration. *If* Fatima is right that she is just playing a bit of pretend, and *if* she is right that, by contrast, Taika *genuinely* enjoys the sadistic murder of innocents in videogames, and this reflects something about his genuine moral character, then it seems appropriate to make a moral evaluation of Taika but not of Fatima.

17 See Bartel (2020) for a fuller defense of virtue ethics in the context of videogames.

18 The VCP borrows its name from the problem of personal identity articulated in Schectman (1996).

19 Frankfurt (1976).

20 Watson (1975).

21 Watson (1987, 150–151).

22 Taylor (1989), MacIntyre (1984), Schechtman (1996).

23 Other examples of actions that are not properly attributable to the cause of the action include actions performed under external influence (e.g. coercion, hypnosis, etc.), unintended reflexes or reactions, and deceptive acts meant to mask the actor's true intentions.

24 I do not mean this to be an analysis of role-playing in games more generally. Rather, it is meant to pick out the phenomenon that sometimes occurs in games where players act based on an adopted role. The source of the role, why it is adopted, etc. will vary from player to player. For a more nuanced account of the various factors involved in role-playing, see Deterding and Zagal (2018).

25 Gervais and Merchant (2006) Many actors will reject this approach to acting. Not being an actor myself, I make no claims about the effectiveness of this technique.

26 It is possible for a griefer to create a fictional background for their in-game character such that the character is mean spirited towards others, and then act on the basis of this fictional role. In such a case, the griefer would be engaging in simple roleplaying.

27 The ceteris paribus clause here is to rule out *other* factors that would prevent the action from being attributed to the player. For example, if the player is being coerced to play a videogame in a particular manner, then this would also prevent the action from being properly attributable to the player.

28 Ruscella and Obeid (2021).

29 Ramirez and Labarge (2018).

30 Slater et al. (2006).

31 Francis et al. (2017). Other studies on this dilemma in VR include Patil et al. (2014) and Navarrete et al. (2012).

32 See Kissel (2020) for more detail on these arguments.

33 We might use a bit of game studies lingo here: the "real" identity is the one outside of the magic circle, while the "roleplayed" identity is the one adopted inside the magic circle. Salen and Zimmerman (2003).

34 Some may take umbrage with the claim that these particular roles conflict. I do not wish to make substantive claims about Catholic doctrine or the LGBTQIA+ community and invite the reader to consider their own example of roles that conflict if the present one fails. A person's role as a parent may conflict with their role as a businessperson by placing opposing demands on them. A person's role as an educator may conflict with their role as a friend. Generally,

these kinds of conflicts are not resolved by jettisoning a role, but by *choosing* what to do on a particular occasion on the basis of the variety of roles they inhabit.

35 Maan (2009) argues that conflict between narrative roles is an essential opportunity for creation of one's self-identity in the first place.

36 There is some debate about whether virtual worlds are truly fictional worlds. For an argument in defense, see Tavinor (2009). For an argument against, see Aarseth (2007). See also their contributions in this volume.

37 Vikaros and Degand (2010).

38 Miller (1980).

39 Taylor (1989); Schechtman (1996); Ricoeur (1991).

40 Schechtman (2012, 335).

41 Schechtman (1996, 96).

42 Strawson (2004).

43 Keep in mind the importance of the *virtual* requirement. The question is not whether pressing the "X" button is attributable to the player. Players may adopt the narrative role of "videogame player" in the trivial sense that they view themselves as a player of a game. For the purposes of moral evaluations, however, the question is whether the action *as it occurs in the game* is attributable to the player.

44 Maloney and Freeman (2020, 515).

45 Maloney and Freeman (2020, 514).

References

Aarseth, E. (2007). Doors and perception: fiction vs. simulation in games. *Intermedialites*, 9: 35–44.

Ali, R. (2015). A new solution to the gamer's dilemma. *Ethics and Information Technology*, 17 (4): 267–274.

Ali, R. (2022). The video gamer's dilemmas. *Ethics and Information Technology*, 24 (2): 1–8. Retrieved from https://doi.org/10.1007/s10676-022-09638-x

Bartel, C. (2012). Resolving the gamer's dilemma. *Ethics and Information Technology*, 14 (1): 11–16.

Bartel, C. (2015). Free will and moral responsibility in video games. *Ethics and Information Technology*, 17, 285–293.

Bartel, C. (2020). *Video games, violence, and the ethics of fantasy*. London: Bloomsbury Academic.

Deterding, S., & Zagal, J. (Eds.) (2018). *Role-playing game studies: Transmedia foundations*. New York: Routledge.

Francis, K. B., Terbeck, S., Briazu, R. A., Haines, A., Gummerum, M., Ganis, G., & Howard, I. S. (2017). Simulating moral actions: an investigation of personal force in virtual moral dilemmas. *Scientific Reports*, 7 (1): 1–11. Retrieved from http://dx.doi.org/10.1038/s41598-017-13909-9

Frankfurt, H. (1976). Identification and externality. In *The identities of persons*. A.O. Rorty (ed). Berkeley: University of California Press: 239–252.

Freeman, G., Zamanifard, S., Maloney, D., & Adkins, A. (2020). My body, my avatar: how people perceive their avatars in social virtual reality. *Conference on Human Factors in Computing Systems - Proceedings*: 1–8.

Gervais, R., & Merchant, S. (2006). *Extras- sir ian mckellen*. UK: BBCTwo.

Goerger, M. (2017). Value, violence, and the ethics of gaming. *Ethics and Information Technology*, 19 (2): 95–105.

Heinrichs, J. H. (2021). Virtual action. *Ethics and Information Technology*, 23: 317–330. Retrieved from https://doi.org/10.1007/s10676-020-09574-8

Hernandez, P. (2019). Steam game about raping women will test valve's hands-off approach. Retrieved February 3, 2023, from https://www.polygon.com/2019/3/4/18249916/rape-day-steam-valve

Kissel, A. (2020). Free will, the self, and video game actions. *Ethics and Information Technology*, 23: 177–183.

Luck, M. (2009). The gamer's dilemma: an analysis of the arguments for the moral distinction between virtual murder and virtual paedophilia. *Ethics and Information Technology*, 11 (1): 31–36.

Luck, M. (2018). Has Ali dissolved the gamer's dilemma? *Ethics and Information Technology*, 20 (3): 157–162.

Maan, A. K. (2009). *Internnarative identity: Placing the self*. University Press of America.

MacIntyre, A. C. (1984). *After virtue: A study in moral theory*. Notre Dame: University of Notre Dame Press.

Maloney, D., & Freeman, G. (2020). Falling asleep together: what makes activities in social virtual reality meaningful to users. *CHI PLAY 2020 - Proceedings of the Annual Symposium on Computer-Human Interaction in Play*: 510–521.

Miller, M. J. (1980). Role-playing as a therapeutic strategy: A research review. *The School Counselor*, 27 (3): 217–226.

Navarrete, C. D., McDonald, M. M., Mott, M. L., & Asher, B. (2012). Virtual morality: emotion and action in a simulated three-dimensional "trolley problem." *Emotion (Washington, D.C.)*, 12 (2): 364–370.

Ostritsch, S. (2017). The amoralist challenge to gaming and the gamer's moral obligation. *Ethics and Information Technology*, 19 (3): 117–128.

Patil, I., Cogoni, C., Zangrando, N., Chittaro, L., & Silani, G. (2014). Affective basis of judgment-behavior discrepancy in virtual experiences of moral dilemmas. *Social Neuroscience*, 9 (1): 94–107.

Patridge, S. (2011). The incorrigible social meaning of video game imagery. *Ethics and Information Technology*, 13 (4): 303–312.

Ramirez, E. J., & LaBarge, S. (2018). Real moral problems in the use of virtual reality. *Ethics and Information Technology*, 20 (4): 249–263. Retrieved from http://dx.doi.org/10.1007/s10676-018-9473-5

Ricoeur, P. (1991). Narrative identity. In *On Paul Ricoeur: Narrative and Interpretation*. D. Wood (ed). London: Routledge.

Ruscella, J. J., & Obeid, M. (2021). A taxonomy for immersive experience design. *7th International Conference of the Immersive Learning Research Network*, 1–5.

Salen, K., & Zimmerman, E. (2003). *Rules of play: Game design fundamentals*. Cambridge: MIT Press.

Schechtman, M. (1996). *The constitution of selves*. Ithaca, NY: Cornell University Press.

Schechtman, M. (2012). The story of my (second) life : virtual worlds and narrative identity. *Philosophy and Technology*, 25 (3): 329–343.

Slater, M., Antley, A., Davison, A., Swapp, D., Guger, C., Barker, C., ... Sanchez-Vives, M. (2006). A virtual reprise of the Stanley milgram obedience experiments. *PLoS One*, 1 (1): e39.

Strawson, G. (2004). Against narritivity. *Ratio*, XVII: 428–452.

Tavinor, G. (2009). *The art of videogames*. Oxford: Wiley- Blackwell.

Taylor, C. (1989). *Sources of the self*. Cambridge, MA: Harvard University Press.

Vikaros, L., & Degand, D. (2010). Moral development through social narratives and game design. In *Ethics and game design: Teaching values through play*. K. Schrier & D. Gibson (eds). IGI Global: 197–215. https://doi.org/10.4018/978-1-61520-845-6.ch013. Retrieved from https://escholarship.org/uc/item/72v253hj

Watson, G. (1975). Free agency. *Journal of Philosophy*, 72 (April): 205–220.

Watson, G. (1987). Free action and free will. *Mind*, 96 (382): 145–172.

Young, G. (2017). Objections to Ostritsch's argument in "the amoralist challenge to gaming and the gamer's moral obligation." *Ethics and Information Technology*, 19 (3): 209–219. Retrieved from http://dx.doi.org/10.1007/s10676-017-9437-1

Part 3

What Can We Do with Extended Realities?

8 Virtual Reality in Experimental Moral Psychology

Identifying and Understanding Judgment-Action Discrepancy

Kathryn B. Francis

Is recycling a moral duty? Is it morally acceptable to throw recyclable products into the waste? Many people would respond "yes" to the first question and "no" to the latter. Yet if I were to inspect their household waste, it is likely that many of these individuals would not be *practising what they preach*. This dissociation between what we believe to be morally right and our actions in ethical contexts has become known as a *moral inconsistency* (Monin & Merritt, 2012) or *moral judgment-action discrepancy* (Francis et al., 2016; Patil, Cogoni, Zangrando, Chittaro, & Silani, 2014). Although some might consider the topic of recycling to be trivial, we continue to see examples of this judgment-action discrepancy when the stakes are raised and when actions can result in life or death. In this chapter, I consider the methodological and ethical challenges of measuring moral actions in the domain of morality of harm, with a focus on the introduction of Virtual Reality (VR) paradigms to study moral inconsistency.

Sacrificial Moral Dilemmas and Models of Moral Decision-Making

For many years, ethicists and moral psychologists have considered the role of moral schools of thought in guiding moral judgments (Doris, 2010). In the classic *trolley problem*, participants are presented with the following case:

> A runaway trolley is heading down a track towards five people who cannot hear it approaching. *You* are standing next to a lever which operates a set of points, or switch, on the track. Pull the lever and the runaway trolley will head down a neighbouring track instead where there is only one individual.

Participants are then asked whether they think it is morally acceptable to flick the switch, sacrificing one life in order to save five others (Thomson, 1976). An affirmative response is considered "characteristically utilitarian" in the sense of maximising the greatest outcome for the greatest number

DOI: 10.4324/9781003359494-12

of people[1]. In a variation of this moral dilemma known as the *footbridge dilemma*, participants are challenged further:

> A runaway trolley is heading down a track towards five people who cannot hear it approaching. *You* are standing on a footbridge in between the people and the oncoming trolley. In front of you stands a very large stranger. Push this stranger off the footbridge and their large bulk will derail the train stopping its course towards the five people.

Again, participants are asked the extent to which they believe the action of sacrificing one life to save many more is morally acceptable (Foot, 1978), and in this case, specifically, the extent to which pushing this person off the footbridge would be morally acceptable. Both philosophers and psychologists have been intrigued by responses to switch-type versus footbridge-type cases, as despite their structural similarity in entailing the one-for-five trade-off, a majority of people are *utilitarian* in switch cases but an extreme minority are in *footbridge* cases (e.g., Foot, 1978; Greene et al., 2001, 2004). In attempts to understand these differences, several models have been proposed.

In Greene's dual process model of moral judgment, a distinction is drawn between *personal* footbridge-type moral dilemmas and *impersonal* switch-type cases (Greene, Sommerville, Nystrom, Darley, & Cohen, 2001). Personal moral dilemmas are those that involve "actions that are (a) likely to cause serious bodily harm, (b) to a particular person, where (c) this harm does not result from deflecting an existing threat into a different party" (Greene et al., 2001, p. 2107). When reading and responding to a personal moral dilemma, the model proposes that we experience an aversive emotional response; an "alarm bell" triggered by emotional systems (Cushman, Young, & Greene, 2010). This aversive reaction drives the non-utilitarian response (e.g., that it is not morally acceptable to push the person off the footbridge). In impersonal moral dilemmas, on the other hand, the actions do not meet the criteria outlined above and so do not trigger the same aversive emotional reactions. The model proposes that in the absence of the alarm bell, the utilitarian decision to sacrifice one life to save many can dominate, driven by increased activations in brain areas associated with cognitive control (Greene et al., 2001).

While the original model proposed by Greene does appear to centre on the antagonism between emotion and reason, this division is somewhat artificial (Greene, Nystrom, Engell, Darley, & Cohen, 2004) and "...overly simple" (Cushman et al., 2010). Removing this arbitrary division, both Cushman (2013) and Crockett (2013) proposed model-based dual-process accounts of moral judgment. In these accounts, one process involving model-free processing assigns a value to an *action* in a moral dilemma (e.g., pushing a person off a footbridge) and another process involving model-based processing selects actions based on a value assigned to their *outcome* instead (e.g., saving the most lives). The disparity in responses to personal footbridge-type dilemmas and impersonal switch-type dilemmas comes from differences in value assignment; a model-based system assigns a positive value to the outcome of saving the most lives in the switch case whereas a model-free system assigns a negative value to the action of pushing (harming) in the footbridge case.

The Measurement of Morality: The Problem with Trolleys and Text-Based Vignettes

While a substantial body of work in moral psychology has considered the differences in responses to personal versus impersonal moral dilemmas, comparatively little research has considered (1) how we present these moral dilemmas to participants and (2) how we measure moral decisions in response to them. For example, research has raised concerns regarding the formulation of these moral dilemmas in terms of word length and framing (Christensen, Flexas, Calabrese, Gut, & Gomila, 2014; Petrinovich & ONeill, 1996) as well as the role of moral principles within these cases (Christensen et al., 2014). For example, individuals are more accepting of harmful actions that will benefit themselves over strangers (the so-called self-beneficial versus other beneficial contrast) (Christensen et al., 2014) and more accepting of harmful actions towards agents who would have died anyway (inevitable harm versus avoidable harm) (e.g., Hauser, 2006). In fact, Kahane et al. (2015) and Bauman et al. (2014) have argued that these sacrificial moral dilemmas tell us very little about *utilitarianism*; the fact that utilitarian responses in these cases are predicted by psychopathic personality traits and further "irrelevant features" gives weight to their concerns (e.g., Bartels & Pizarro, 2011; Djeriouat & Tremoliere, 2014; Kahane et al., 2015; Patil, 2015).

Of particular relevance here is the work of Tassy, Oullier, Mancini, and Wicker (2013) who found that even the question that participants were asked after reading a moral dilemma, impacted their moral judgments. When asked an action-choice question ("Would you do it?") participants gave more utilitarian responses than when asked a judgment question ("Is it morally acceptable?"). When applying Repetitive Transcranial Magnetic Stimulation (rTMS)[2] to an area of the brain associated with executive control functions (the dorsolateral prefrontal cortex), Tassy et al. (2012) found that moral judgments ("Is it morally acceptable?") were affected but not action-choices ("Would you do it?"). The researchers concluded that these results support a judgment-action discrepancy whereby moral actions and moral judgments are driven by partially distinct mechanisms and are thus, distinct constructs (Camerer & Mobbs, 2017; Tassy, Deruelle, Mancini, Leistedt, & Wicker, 2013; Tassy, Oullier et al., 2013). This also supports research showing discrepancies between what people say and do in other ethically-relevant contexts (e.g., Olafson, Schraw, Nadelson, Nadelson, & Kehrwald, 2013).

However, the extent to which an action-choice question ("Would you do it?") can accurately measure moral action is debated. These questions remain *hypothetical* and even Tassy et al. (2013) state that responses to these are "...obviously what the participants think their action could be if they were to make the decision in real life" (Tassy, Oullier et al., 2013, p. 2). These text-based vignettes offer limited immersion and lack contextual information (Skulmowski, Bunge, Kaspar, & Pipa, 2014) and in fact, research has found evidence that increasing the amount of contextual information in a scenario can bring hypothetical moral decisions in line with real moral decisions (see FeldmanHall et al., 2012; Gold, Pulford, & Colman, 2014). As such, while the use and formulation of text-based sacrificial moral dilemmas are valuable in shedding light on the processes that underlie moral judgment (Christensen & Gomila, 2012), we cannot determine how hypothetical responses in these contextually-impoverished paradigms would translate into tangible moral actions in environments that are free from the problems of de-contextualisation (Francis et al., 2016; Parsons, 2015).

Virtual Morality: Moral Actions in Contextually Rich Moral Dilemmas

Despite the limitations of utilising text-based sacrificial moral dilemmas to understand moral judgments and particularly moral actions, there are two principal challenges in designing methodologies that can assess moral actions in the domain of morality of harm. Firstly, and given the life-and-death nature of the scenarios, we cannot expose participants to these situations in real life to observe and measure their moral actions (e.g., Patil et al., 2018). Secondly and given that these sacrificial moral dilemmas are often far-fetched and extreme in nature, it is unlikely that we would find such occurrences of them in the real world (e.g., Bauman et al., 2014). Virtual Reality (VR) simulations provide a reasonable solution to both the ethical challenge and the methodological challenge above. Using VR, we are able to simulate impossible worlds and events that would not occur in real life (e.g., Riva, Banos, Botella, Mantovani, & Gaggioli, 2016) and we can also protect the physical welfare of our participants (Francis et al., 2016). There is now a considerably large body of research evidencing the methodological value of VR paradigms across various research settings in offering a balance between experimental control, experimental realism, and ecological validity (e.g., Parsons, 2015; Parsons, Gaggioli, & Riva, 2017).[3] It is also worth noting that VR simulations support the researcher in observing and recording human action *in the moment* of decision-making and so in a virtual simulation of a sacrificial moral dilemma, the researcher can record (1) the moral action as it happens in (2) a real-time sequence of events and in (3) a contextually salient environment.

In early research in this area, Navarrete, McDonald, Mott, and Asher (2012) created a virtual simulation of the switch dilemma. Comparing simulated actions in the VR dilemma to hypothetical moral judgments in a text-based counterpart, the researchers found that most people gave characteristically utilitarian responses in *both* modes of the impersonal moral dilemma. Going back to models of moral judgment, the researchers argue that the high utilitarian responses seen in the VR simulation, suggest that Greene's dual process model of moral judgment can be generalised to the context of moral actions as it explains the utilitarian endorsement pattern in both the text-based and VR impersonal switch dilemma (Greene et al., 2001; Navarrete et al., 2012). Skulmowski et al. (2014) extended this work, creating a VR switch dilemma in which the user became the driver of the train car. Again, response patterns in the VR simulation were predominately utilitarian supporting Greene's predictions that in switch-type impersonal cases, controlled cognitive processes drive the decision-making in favour of maximising the greatest outcome for the greatest number of people.

However, not all VR studies in the field have found this clear-cut consistency between moral judgments made in responses to sacrificial moral dilemmas and their VR counterparts (e.g., Pan & Slater, 2011). For example, Patil et al. (2014) found that third-person virtual switch-type moral dilemmas elicited more utilitarian actions when compared to moral judgments in their text-based counterparts. The immersive nature of VR appears to play a role here too as Pan and Slater (2011) found consistent responses between text-based vignettes and VR actions generally but greater utilitarian responses in immersive VR moral dilemmas when compared to non-immersive VR counterparts.[4]

Later research in the field has been able to examine moral principles such as action, intention, contact, and personal force[5] in VR by developing simulations of up-close and personal footbridge-type dilemmas instead (e.g., Francis et al., 2016) (see Figure 8.1). For example, when comparing simulated actions in an immersive VR version of the footbridge dilemma to moral judgments made in response to a text-based counterpart, Francis et al. (2016) found substantially greater utilitarian endorsements in VR (see Figure 8.2).

Figure 8.1 A virtual reality simulation of the footbridge dilemma.

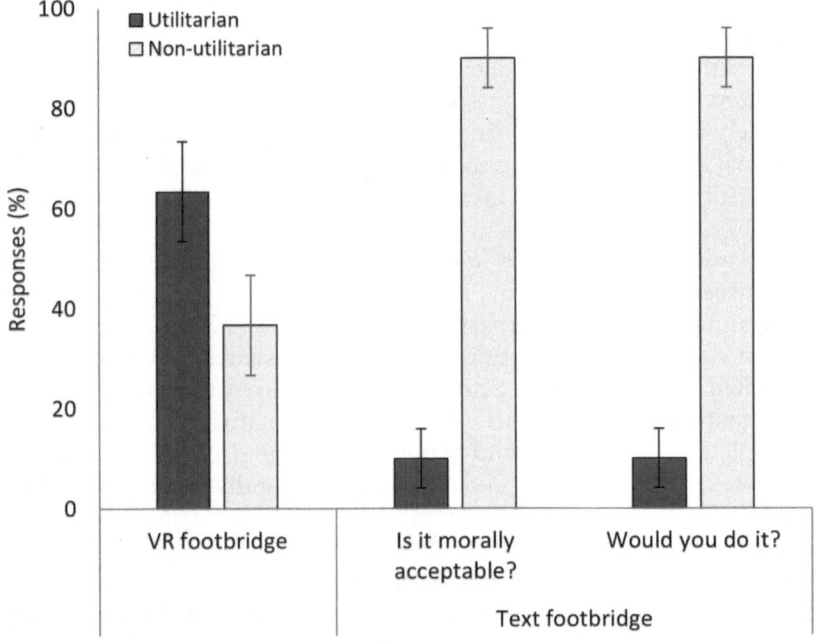

Figure 8.2 Responses to the VR footbridge dilemma versus the text-based counterpart presented in Francis et al. (2016). Participants responded to
both a judgment and action-choice question when responding to the
text-based version of the dilemma. Participants gave greater utilitarian
responses (via a joystick device) when simulated action was required in
VR. Error bars represented +−1 standard error.

The VR environment allows the user to experience a moral dilemma in a 360° from the first-person perspective. All features of the VR scene can be designed, adapted, and controlled by the experimenter including the agents' appearances, the skybox, and the terrain. Users can interact with the environment using input devices such as joystick devices or hand controller.

In Francis et al. (2017), the VR paradigm was further developed to account for the moral principles of personal force and physical contact (e.g., see Christensen et al., 2014; Greene et al., 2009). In study one, a robotic haptic VR interface (vBOT) system was introduced so that if participants chose to endorse a utilitarian response, they would have to push a robotic arm forward generating the physical resistance force that would be experienced if making contact with the object described in the scenario. For example, if participants wanted to push the person off the bridge in the footbridge dilemma, they would push the vBOT arm forward and they would experience haptic feedback as if they had pushed another person (see Figure 8.3a). In study two, corporeal sensations were also incorporated by creating an interactive sculpture designed to act as the response device in the VR footbridge dilemma. The response device was a person's torso built using platinum-grade silicon, heated wiring, and was weighted to simulate a realistic body weight (see Figure 8.3b). The scene was set up so that the physical torso aligned with the virtual agent in VR. If the participants wanted to push the virtual person off the footbridge in VR, they would have to push this counterpart's torso in front of them[6]. In both studies, utilitarian responses were greater in the virtual paradigms even when harmful actions were made to feel more realistic via haptic feedback and realistic touch. This pattern of greater utilitarian endorsements in VR and haptic-VR simulations when compared to text has been found in further studies in the field since (see Francis, Gummerum, Ganis, Howard, & Terbeck, 2018, 2019; McDonald, Defever, & Navarrete, 2017).

Moral Judgment-Action Discrepancy and Mode of Presentation

According to Greene's dual process model of moral judgment, we can explain low utilitarian endorsements in footbridge-type moral dilemmas across traditional paradigms (text-based) because these *personal* moral dilemmas trigger immediate and aversive emotional responses driven by emotional processing centres in the brain (Greene et al., 2001). But can this model explain the responses in VR simulations of these moral dilemmas? Given that *personal* moral dilemmas prompt emotionally-driven responses, and the fact that the immersive nature of VR triggers even greater emotional arousal (e.g., Parsons & Rizzo, 2008), the model would predict

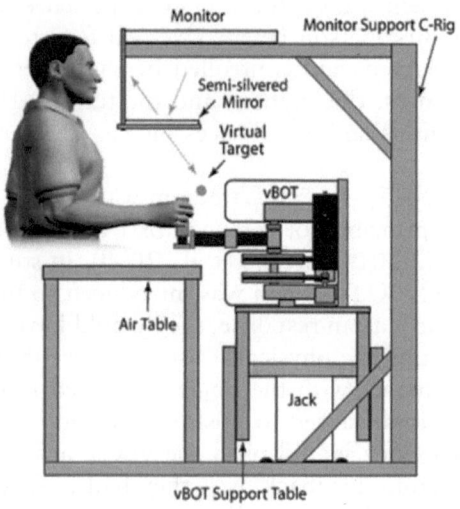

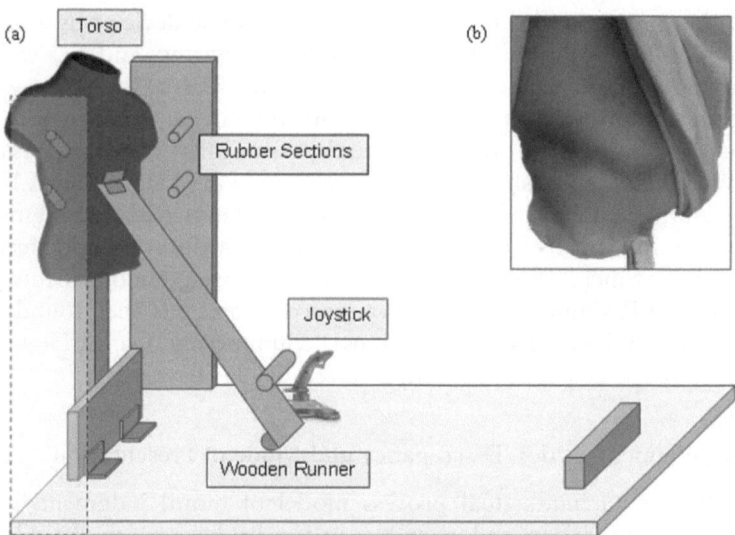

Figure 8.3 Virtual and haptic systems utilised in Francis et al. (2017) (a) The participant holds the handle of the vBOT arm and responds to the moral dilemma by either pushing forward to endorse the utilitarian outcome or pulling towards themselves to endorse a nonutilitarian outcome (b) The interactive sculpture in study two resembled a person's torso and the sculpture would generate resistance if pushed. A wooden runner captured the weight of the torso as it fell which would trigger the pushing of the virtual counterpart in the VR simulation (a). The torso was built using platinum-grade silicon with heated wiring to generate a body-like temperature (b).

even *fewer* utilitarian endorsements in virtual moral dilemmas. However, several studies in the field have found the opposite with approximately 50–70% utilitarian response rates in virtual personal moral dilemmas (Francis et al., 2016, 2017, 2018, 2019; McDonald et al., 2017). Existing research has claimed that this reflects a *judgment-action discrepancy*; VR supports the measurement of moral action by immersing participants in an environment in which an egocentric perspective drives decision-making as users consider the consequences of their own actions unlike a text-based paradigm in which allocentric perspective taking (what other people and society might expect us to do) drives non-utilitarian decision-making (Francis et al., 2017; Tassy, Oullier et al., 2013).

Further, both Patil et al. (2014) and Francis et al. (2016) have argued that an adapted version of Cushman (2013)s and Crockett (2013)s reinforcement learning models offer a plausible explanation of this surprising response pattern in VR. In a virtual simulation of a moral dilemma, the user is no longer relying on imagination and the process of making the one-life-for-five-lives comparison is made both visually salient and fast. As such, a greater negative value might be assigned to the outcome of seeing five people die and this may outweigh the negative value assigned to the action of harming one individual, thus tipping the decision in favour of the utilitarian outcome.

Whether the current findings from VR research in the field reflect a *true* judgment-action discrepancy has yet to be determined but there is increasing evidence that these findings reflect something beyond a simple effect of mode of presentation (text versus VR). For example, in recent work, we have shown that responses to moral dilemmas do not differ when scenarios are presented in text format compared to images or when text-based vignettes are compared to moral dilemmas constructed using images with sounds (Francis et al., *in prep*). Further, responses to moral dilemmas do not appear to be influenced by the inclusion of additional contextual features such as moving image sequences of a threat with sounds (such as a moving trolley) or 2D videos of the moral dilemmas accompanied by sounds (such as a first-person video of the footbridge dilemma) (Francis et al., *in prep*). This suggests that increasing the contextual richness of the scenario using mode of presentation is not the *only* factor contributing to the greater endorsement of utilitarian outcomes in VR moral dilemmas.

Perhaps more importantly, research has investigated this apparent moral judgment-action discrepancy beyond extreme trolley-type cases. For example, Patil et al. (2018) created a virtual simulation in which participants were tasked with escaping a burning building. During their escape, users come across an injured individual, and they are faced with a dilemma; do they stop to help the injured person but at a risk to their own life, or do they leave this injured person behind and save themselves? In a text-based vignette version of this altruistic moral dilemma, a majority of participants stated that they would *save* the injured person even if it would risk their own life (91%) and that this was morally permissible. However, in the virtual simulation, fewer participants were altruistic (65%) and more chose to leave and *sacrifice* the injured person to save themselves. The authors argue that responses to text-based vignettes can be "…notoriously unreliable for accurately predicting…actual behaviour" with VR offering opportunities to directly address this (Patil et al., 2018, p. 35). Thus we see evidence of a moral judgment-action discrepancy in virtual simulations other than sacrificial moral dilemmas.

Of course, any virtual simulation is limited regardless of plausibility in that there are no tangible consequences for the actor (Bostyn, Sevenhant, & Roets, 2018; Patil et al., 2018). To address this and to move beyond VR simulations altogether, Bostyn et al. (2018) invited participants to the lab where they were shown two boxes; one box contained five live mice and the other contained one live mouse. Participants were told that live wiring connected to the boxes would administer a painful but nonlethal shock to the five mice unless redirected to the box with a single mouse via a button press[7]. In the lab-based behaviour task, a majority of participants redirected the electric shock endorsing the utilitarian outcome (84%). However, responses to a text-based vignette version of the same situation produced contrasting responses with far fewer participants stating that they would *press the button,* endorsing a utilitarian outcome (66%). The authors argue that their findings corroborate those of VR studies in the field that find higher rates of utilitarian endorsement when "…confronting participants with dilemmas that are more lifelike" (Bostyn et al., 2018, p. 1090). This study provides evidence that (1) a moral judgment-action discrepancy does exist and (2) that findings in VR simulations in the field substantiate those seen in a behavioural task where the consequences are both real and tangible.

Moral Hypocrisy, Akrasia, and VR Trolley Problems

Bertrand Russell claimed that we have two forms of morality running in parallel: "one which we preach but do not practice and another which we practice but seldom preach" (Russell, 2004). This *moral hypocrisy*, when individuals accept moral norms and make appropriate moral judgments, but then fail to act in line with these (Batson et al., 1997) is something that each of us is likely guilty of in one or more aspects of our everyday lives. In philosophy, moral hypocrisy can be linked to *akrasia*, the notion of acting against our better judgment due to a lack of control or weakness of will (see Bobonich & Destrée, 2007). To what extent does the moral judgment-action discrepancy evidenced in VR (and related research) in this field connect to these concepts?

Firstly, we need to raise an important distinction between examining the relationship between moral judgment and moral action versus examining moral judgments and implications for action in the predictive sense. VR research published in the domain of morality of harm has often adopted a between-subjects design whereby *different* participants respond to text-based trolley-type problems and VR-based trolley-type problems (e.g., Francis et al., 2016, 2017). These studies can shed light on the relationship between moral judgments in response to text-based vignettes versus moral actions simulated in VR counterparts, supporting a judgment-action discrepancy. However, given that different people provide moral judgments and moral actions, we are unable to examine moral judgments and implications for action, and so subsequently, we cannot investigate moral hypocrisy and/or akrasia directly.

However and in some cases, VR studies in the field have adopted within-subjects whereby the *same* participants respond to text-based moral dilemmas and their VR counterparts (e.g., Francis et al., 2019; Patil et al., 2014). In Francis et al. (2019) for example, participants were presented with the footbridge dilemma in VR *and* a text-based counterpart dilemma (matched for the moral principles of benefit-recipient, inevitability, moral magnitude, physical contact, and personal force). Participants demonstrated the same disparity in responses to VR moral dilemmas versus text-based dilemmas as seen in previous studies using between-subjects designs; participants were more likely to simulate utilitarian actions in VR but refused to endorse utilitarian responses when asked a judgment and action-choice question. In this study, the order of tasks was counterbalanced meaning that half of the participants received the text-based moral dilemma and judgment questions first and the other half received the VR moral dilemma requiring action first. Arguably, this study allows the investigation of moral hypocrisy and akrasia in the participants who were required to make a judgment first and an action second, allowing the direct examination of the extent to which formed moral judgments predict subsequent moral actions. Taking this view when interpreting the results of within-subjects studies, we might conclude that people do demonstrate moral hypocrisy by *judging* in one way but subsequently *acting* in another. Nevertheless, there has yet to be an explicit investigation of this moral hypocrisy or akrasia and the extent to which lack of control or weakness of will are responsible for the disparity in responses between text and VR trolley-type problems presented in the field.

The Future of Virtual Reality in Moral Psychology

Virtual Reality has created an opportunity to study simulated moral actions within the domain of morality of harm, something that has been both ethically and methodologically impossible previously. There is now a body of research demonstrating that moral responses in these contextually rich and simulated environments (1) differ from moral decisions given in response to traditional text-based vignettes (Francis et al., 2016, 2017, 2018, 2019; McDonald et al., 2017; Patil et al., 2014, 2018) and that (2) these may represent moral *actions* rather than moral judgments (Francis et al., 2016; Patil et al., 2014). However, the extent to which VR can be used to investigate and measure the underlying construct of moral action in the domain of morality of harm requires further investigation. The ultimate assessment of the ecological validity of this body of VR research would be to determine if utilitarian endorsements in VR map onto utilitarian actions in real life. Although ethically and practically challenging, VR-simulated responses could be compared to behaviours in real-life moral dilemmas such as applied triage decisions.

Further, the work emerging in this domain has supported the introduction of new and adapted models of moral decision-making that explain (1) why moral responses differ between VR simulations and text-based vignettes via differences in the assignment of values to actions versus outcomes (Francis et al., 2016; McDonald et al., 2017; Patil et al., 2014) and (2) why moral actions and moral judgments differ based on opposing perspectives driving decision-making (Francis et al., 2017; Tassy, Oullier et al., 2013). Of course, existing models of moral decision-making were created to explain moral *judgment* specifically and further research is required to determine the extent to which these accounts can explain VR findings. For example, collecting eye-tracking data in VR simulations could provide information regarding where attention is placed in the scene (the five potential victims or the one individual) and this data could determine who participants are most concerned about and thus, where value is being assigned. However, while some research has found that longer gaze durations predict who will be harmed in a moral dilemma (Skulmowski et al., 2014), there are mixed findings on this with some studies showing that people tend to look away from targeted victims (Kastner, 2010). Further, regarding the role of perspectives on decision-making, future research might consider manipulating allocentric versus egocentric decision-making in these virtual simulations (e.g., by introducing third-party bystanders).

With Extended Reality (XR) technologies advancing rapidly, we can anticipate the release of increasingly sophisticated virtual technologies that will support the generation of realistic-looking and feeling simulations. This is promising for moral psychology and for future investigations of moral actions that are guided by certain moral schools of thought. Creating simulations that have both perspectival fidelity (accurate representation of the subjective point-of-view of the user) and contextual realism (accurate representation of the content of the experience) can produce virtually real experiences (Ramirez & LaBarge, 2018, 2020) and this will be facilitated by improvements in the resolution of VR simulations and the quality of response devices. These advancements will support the investigation of realistic simulated moral actions that will be progressively more likely to predict real-life behaviours in similar situations. However and despite the obvious advantages, moral psychologists should proceed with caution. While VR moral dilemmas do not cause *physical* harm to participants, Ramirez and LaBarge (2018) argue that "if it would be wrong to allow subjects to have a certain experience in reality, then it would be wrong to allow subjects to have that experience in a virtually real setting" (p. 249). Should a vivid and salient virtually-constructed moral dilemma prompt the same emotional reaction and conflict in a participant that might be experienced in a real counterpart of the same dilemma, we should pause to consider the risk and impact of such a simulation. Given that virtual simulations are now used across helping professions that face challenging triage decisions on a regular basis (such as paramedics and fire service professionals), appropriate briefing and debriefing criteria alongside updated ethical procedures will need to be developed. If researchers can establish ethical boundaries for VR investigations of moral actions however, XR technologies will continue to play a critical role in the future of moral psychology, offering unique opportunities to investigate harmful moral actions in up-close and personal scenarios, in ways that have not been previously possible.

Notes

1 Note that Greene (2008) uses the term *characteristically* in a meaningful way here as it is not likely that the action of choosing to flick the switch (or similar) captures or reflects a complete picture of utilitarian ideals (e.g., Bauman, McGraw, Bartels, & Warren, 2014; Kahane, Everett, Earp, Farias, & Savulescu, 2015)
2 rTMS uses multiple magnetic pulses to either stimulate (high-frequency rTMS) or inhibit (low-frequency rTMS) targeted areas of the cerebral cortex.
3 This has value for experimental research as typically, the more the researcher attempts to control the study to remove the effects of extraneous variables (experimental control), the less likely the findings from the contrived lab or study environment will be generalisable to the real-world (ecological validity) and vice versa.
4 The results of this study should be interpreted tentatively as this was a pilot with a small sample size (n = 36).
5 "...when the force that directly impacts the other [person] is generated by the agent's muscles" (Greene, 2009, p. 365). For example, directly pushing someone would require personal force but firing a gun at someone would not.
6 For full details, please see Francis et al. (2017).
7 Note that no real shock was administered but that participants were deceived into believing so.

References

Bartels, D. M., & Pizarro, D. A. (2011). The mismeasure of morals: Antisocial personality traits predict utilitarian responses to moral dilemmas. *Cognition, 121*(1), 154–161. Doi:10.1016/j.cognition.2011.05.010

Batson, C. D., Kobrynowicz, D., Dinnerstein, J. L., Kampf, H. C., & Wilson, A. D. (1997). In a very different voice: unmasking moral hypocrisy. *Journal of Personality and Social Psychology, 72*(6), 1335.

Bauman, C. W., McGraw, A. P., Bartels, D. M., & Warren, C. (2014). Revisiting external validity: Concerns about trolley problems and other sacrificial dilemmas in moral psychology. *Social and Personality Psychology Compass, 8*, 536–554. Doi: 10.1111/spc3.12131

Bobonich, C., & Destrée, P. (Eds.). (2007). *Akrasia in Greek philosophy: From Socrates to Plotinus.* The Netherlands: Brill.

Bostyn, D. H., Sevenhant, S., & Roets, A. (2018). Of mice, men, and trolleys: Hypothetical judgment versus real-life behaviour in trolley-style moral dilemmas. *Psychological Science, 29*(7), 1084–1093. Doi:10.1177/0956797617752640

Camerer, C., & Mobbs, D. (2017). Differences in behaviour and brain activity during hypothetical and real choices. *Trends Cognitive Science, 21*(1), 46–56. Doi:10.1016/j.tics.2016.11.001

Christensen, J. F., Flexas, A., Calabrese, M., Gut, N. K., & Gomila, A. (2014). Moral judgment reloaded: A moral dilemma validation study. *Frontiers in Psychology, 5*, 607. Doi:10.3389/fpsyg.2014.00607

Christensen, J. F., & Gomila, A. (2012). Moral dilemmas in the cognitive neuroscience of moral decision-making: A principled review. *Neuroscience and Biobehaviour Reviews, 36*(4), 1249–1264. Doi:10.1016/j.neubiorev.2012.02.008

Crockett, M. J. (2013). Models of morality. *Trends Cognitive Science, 17*(8), 363–366. Doi:10.1016/j.tics.2013.06.005

Cushman, F. (2013). Action, outcome, and value: A dual-system framework for morality. *Personality and Social Psychology Reviews, 17*, 273–292. Doi:10.1177/1088868313495594

Cushman, F., Young, L., & Greene, J. (2010). Our multi-system moral psychology: Towards a consensus view. In J. M. Doris (Ed.), *The Moral Psychology Handbook* (pp. 47–71). Oxford: Oxford University Press.

Djeriouat, H., & Tremoliere, B. (2014). The dark triad of personality and utilitarian moral judgment: The mediating role of honesty/humility and harm/care. *Personality and Individual Differences, 67*, 11–16. Doi:10.1016/j.paid.2013.12.026

Doris, J. M. (2010). Introduction. In J. M. Doris (Ed.), *The Moral Psychology Handbook* (pp. 1–2). Oxford: Oxford University Press.

FeldmanHall, O., Mobbs, D., Evans, D., Hiscox, L., Navrady, L., & Dalgleish, T. (2012). What we say and what we do: The relationship between real and hypothetical moral choices. *Cognition, 123*, 434–441. Doi:10.1016/j.cognition.2012.02.001

Foot, P. (1978). Virtues and vices and other essays in moral philosophy. University of California Press, Berkeley and Los Angeles.

Francis, K. B., Gummerum, M., Ganis, G., Howard, I. S., & Terbeck, S. (2018). Virtual morality in the helping professions: Simulated action and resilience. *British Journal of Psychology, 109*(3), 442–465. Doi:10.1111/bjop.12276

Francis, K. B., Gummerum, M., Ganis, G., Howard, I. S., & Terbeck, S. (2019). Alcohol, empathy, and morality: acute effects of alcohol consumption on affective empathy and moral decision-making. *Psychopharmacology (Berl), 236*(12), 3477–3496. Doi:10.1007/s00213-019-05314-z

Francis, K. B., Howard, C., Howard, I. S., Gummerum, M., Ganis, G., Anderson, G., & Terbeck, S. (2016). Virtual morality: Transitioning from moral judgment to moral action? *PLoS One, 11*(10), 1–22. Doi:10.1371/journal.pone.0164374

Francis, K. B., Terbeck, S., Briazu, R. A., Haines, A., Gummerum, M., Ganis, G., & Howard, I. S. (2017). Simulating moral actions: An investigation of personal force in virtual moral dilemmas. *Scientific Reports, 7*(1), 13954. Doi:10.1038/s41598-017-13909-9

Gold, N., Pulford, B. D., & Colman, A. M. (2014). The outlandish, the realistic, and the real: contextual manipulation and agent role effects in trolley problems. *Frontiers in Psychology, 5*, 35. Doi:10.3389/fpsyg.2014.00035

Greene, J. D. (2008). The secret joke of Kant's soul. In W. Sinnott-Armstrong (Ed.), *Moral Psychology* (Vol. 3, pp. 35–79). Cambridge, MA: The MIT Press.

Greene, J. D. (2009). Dual-process morality and the personal/impersonal distinction: A reply to McGuire, Langdon, Coltheart, and Mackenzie. *Journal of Experimental Social Psychology, 45*(3), 581–584. Doi:10.1016/j.jesp.2009.01.003

Greene, J. D., Cushman, F. A., Stewart, L. E., Lowenberg, K., Nystrom, L. E., & Cohen, J. D. (2009). Pushing moral buttons: the interaction between personal force and intention in moral judgment. *Cognition, 111*, 364–371. Doi:10.1016/j.cognition.2009.02.001

Greene, J. D., Nystrom, L. E., Engell, A. D., Darley, J. M., & Cohen, J. D. (2004). The neural bases of cognitive conflict and control in moral judgment. *Neuron, 44*(2), 389–400. Doi:10.1016/j.neuron.2004.09.027

Greene, J. D., Sommerville, R. B., Nystrom, L. E., Darley, J. M., & Cohen, J. D. (2001). An fMRI investigation of emotional engagement in moral judgment. *Science, 293*, 2105–2108. Doi:10.1126/science.1062872

Hauser, M. (2006). *Moral Minds: How Nature Designed Our Universal Sense of Right and Wrong.* Ecco/HarperCollins Publishers, New York.

Kahane, G., Everett, J. A., Earp, B. D., Farias, M., & Savulescu, J. (2015). 'Utilitarian' judgments in sacrificial moral dilemmas do not reflect impartial concern for the greater good. *Cognition, 134*, 193–209. Doi:10.1016/j.cognition.2014.10.005

Kastner, R. M. (2010). Moral judgments and visual attention: An eye-tracking investigation. *Chrestomathy: Annual Review of Undergraduate Research, School of Humanities and Social Sciences, School of Languages, Cultures, and World Affair, 9*, 114–128.

McDonald, M. M., Defever, A. M., & Navarrete, C. D. (2017). Killing for the greater good: Impersonal versus personal distinctions influence emotional inhibition of harmful action in a virtual moral dilemma. *Evolution and Human Behavior, 38*, 770–778. Doi:doi: 10.1016/j.evolhumbehav.2017.06.001

Monin, B., & Merritt, A. (2012). Moral hypocrisy, moral inconsistency, and the struggle for moral integrity. In P. R. S. M. Mikulincer (Ed.), *The Social Psychology of Morality: Exploring the Causes of Good and Evil, Herzliya Series on Personality and Social Psychology* (Vol. 3, pp. 167–184). Washington, DC: American Psychological Association.

Navarrete, C. D., McDonald, M. M., Mott, M. L., & Asher, B. (2012). Virtual morality: emotion and action in a simulated three-dimensional "trolley problem". *Emotion, 12,* 364–370. Doi:10.1037/a0025561

Olafson, L., Schraw, G., Nadelson, L., Nadelson, S., & Kehrwald, N. (2013). Exploring the judgment-action gap: College students and academic dishonesty. *Ethics and Behavior, 23*(2), 148–162. Doi:10.1080/10508422.2012.714247

Pan, X., & Slater, M. (2011). *Confronting a Moral Dilemma in Virtual Reality: A Pilot Study.* Paper presented at the Proceedings of the 25th BCS Conference on Human-Computer Interaction.

Parsons, T. D. (2015). Virtual reality for enhanced ecological validity and experimental control in the clinical, affective and social neurosciences. *Frontiers in Human Neuroscience, 9,* 660. Doi:10.3389/fnhum.2015.00660

Parsons, T. D., Gaggioli, A., & Riva, G. (2017). Virtual reality for research in social neuroscience. *Brain Science, 7*(4), 42. Doi:10.3390/brainsci7040042

Parsons, T. D., & Rizzo, A. A. (2008). Affective outcomes of virtual reality exposure therapy for anxiety and specific phobias: a meta-analysis. *Journal of Behavior Therapy and Experimental Psychiatry, 39*(3), 250–261. Doi:10.1016/j.jbtep.2007.07.007

Patil, I. (2015). Trait psychopathy and utilitarian moral judgement: The mediating role of action aversion. *Journal of Cognitive Psychology, 27,* 349–366. Doi:10.1080/20445911.2015.1004334

Patil, I., Cogoni, C., Zangrando, N., Chittaro, L., & Silani, G. (2014). Affective basis of judgment-behaviour discrepancy in virtual experiences of moral dilemmas. *Society for Neuroscience, 9,* 94–107. Doi:10.1080/17470919.2013.870091

Patil, I., Zanon, M., Novembre, G., Zangrando, N., Chittaro, L., & Silani, G. (2018). Neuroanatomical basis of concern-based altruism in virtual environment. *Neuropsychologia, 116*(Pt A), 34–43. Doi:10.1016/j.neuropsychologia.2017.02.015

Petrinovich, L., & ONeill, P. (1996). Influence of wording and framing effects on moral intuitions. *Ethology and Sociobiology, 17*(3), 145–171. Retrieved from <Go to ISI>://WOS:A1996UZ39100001

Ramirez, E. J., & LaBarge, S. (2018). Real moral problems in the use of virtual reality. *Ethics and Information Technology, 20,* 249–263. Doi:10.1007/s10676-018-9473-5

Ramirez, E. J., & LaBarge, S. (2020). Ethical issues with simulating the bridge problem in VR. *Science and Engineering Ethics, 26*(6), 3313–3331. Doi:10.1007/s11948-020-00267-5

Riva, G., Banos, R. M., Botella, C., Mantovani, F., & Gaggioli, A. (2016). Transforming experience: The potential of augmented reality and virtual reality for enhancing personal and clinical change. *Frontiers in Psychiatry, 7,* 164. Doi:10.3389/fpsyt.2016.00164

Russell, B. (2004). *Sceptical Essays.* United Kingdom: Routledge.

Skulmowski, A., Bunge, A., Kaspar, K., & Pipa, G. (2014). Forced-choice decision-making in modified trolley dilemma situations: A virtual reality and eye tracking study. *Frontiers in Behavioural Neuroscience, 8,* 426. Doi:10.3389/fnbeh.2014.00426

Tassy, S., Deruelle, C., Mancini, J., Leistedt, S., & Wicker, B. (2013). High levels of psychopathic traits alter moral choice but not moral judgment. *Frontiers in Human Neuroscience, 4*(7), 229. Doi:10.3389/Fnhum.2013.00229

Tassy, S., Oullier, O., Duclos, Y., Coulon, O., Mancini, J., Deruelle, C., ... Wicker, B. (2012). Disrupting the right prefrontal cortex alters moral judgement. *Social Cognitive and Affective Neuroscience, 7*(3), 282–288. Doi:10.1093/scan/nsr008

Tassy, S., Oullier, O., Mancini, J., & Wicker, B. (2013). Discrepancies between judgment and choice of action in moral dilemmas. *Frontiers in Psychology, 4,* 250. Doi:10.3389/Fpsyg.2013.00250

Thomson, J. J. (1976). Killing, letting die, and the trolley problem. *Monist, 59,* 204–217. Retrieved from http://www.ncbi.nlm.nih.gov/pubmed/11662247

9 Moral Behavior in Virtual Reality

Eugy Han and Jeremy N. Bailenson

Research areas surrounding morality evolve over time, given that social, cultural, and political contexts under which these behaviors arize matter. And as societies evolve and new landscapes emerge, it is critical to ask these questions within these contexts. One such context to consider is the virtual world, more specifically social worlds accessed through virtual reality (VR) headsets. Much like how internet communities cultivate their own norms and cultures, it can be expected that the same will apply to worlds within VR. It is especially important that we consider what these norms and cultures look like within social VR given the increasing popularity and interest in the "metaverse,"[1] as well as the unique affordances that VR provides.

The metaverse was first introduced as a fictitious concept in science fiction novels, such as Neal Stephenson's *Snow Crash* (1992), in which the term was first coined. The metaverse can be best described as a virtual world pieced together by digital landscapes and inhabited by digital beings. However, this idea of technology enhancing who and where we are has been explored by writers before then, as seen in novels such as William Gibson's *Neuromancer* (1984), Vernor Vinge's *True Names* (1981), or Philip K. Dick's *The Days of Perky Pat* (1990). What such seemingly distant but increasingly reasonable societies share in common is the need to reexamine how people's behaviors change as core concepts in life – such as the body, our memories, and our personalities – are transformed.

And while it is certainly not reasonable or productive to induce fear of our society heading in the direction of the worlds depicted in such works of fiction, it should be noted that some of the world's most influential shapers of technology are heavily investing their resources and attention to developing their own metaverses. Consequently, there is a growing need to understand how people's moral behaviors change in such environments. As the technology does not exist yet to support the fully realized version of the metaverse, how it is built can be informed by what we can understand through research and history.

DOI: 10.4324/9781003359494-13

A common misconception of the metaverse is that it is intended to be a somewhat veridical representation of the physical world. Experiences within the metaverse are frequently dismissed as not being photographically or behaviorally real enough. However, it is far from veridical. Research on how the affordances of virtual environments, such as what you look like, where you are, or who you are with, affect the way you perceive yourself and behave. We can already see examples of new practices, norms, and interactions taking place – ones that have been shaped by the affordances of different social VR worlds. One such example comes from the spatial property of VR. Many social VR platforms can track users' full-body movement and allow for six degrees of freedom to move around physically. This physicality is novel compared to what can be done in traditional social media or 2D virtual worlds. This affordance has bred novel forms of practices, such as users running around, shouting, and blocking others' views as a way to harass others (Freeman et al., 2022). Another example of a norm found within social VR is what Zheng and colleagues (2023) describe as *immersive dwellers* (or *mirror dwellers*), who are individuals in expensive avatars that spend hours staring at themselves in front of mirrors in public spaces. Some dwellers even sleep inside social VR or leave their headsets while their avatars are "sleeping," inviting other users to harass the sleeping avatars (i.e., by stepping on their faces).

In this chapter, we discuss what research reveals about our behavior within VR. We cover what's been done, what the findings reveal about the metaverse and its participants, and where to go from here. We explore two components that make up the metaverse: the people and the environment. First, the people: how does what you look like transform who you are? We consider how changes in *your* and *others'* visual appearance can lead to changes in behavior and beliefs. Second, the environment: how does virtual reality differ from physical reality? What is the technology capable of and how does this contribute to the ways in which we perceive and experience? Lastly, we finish with the limitations of the current literature and what future research is necessary. Taking a step back, this chapter discusses the new moral questions and scenarios that arize almost uniquely within VR, and why it is critical to evaluate the age-old questions on moral behavior within this new context.

Virtual People: Avatars

You are what you wear. Annie is getting ready for a fancy dinner party. She is deciding on what to wear and how to style her appearance. As much as she would like to show up in loungewear, it would most likely not be appropriate for this context. She also knows that how she presents herself is not only for the other attendees at the party, but for herself as well.

She wants to feel confident. She finds herself correcting her hunched back, parting her hair in different ways till she finds that makes her feel the best.

A few weeks later, Annie is getting ready for another dinner party. Except, this time it is hosted in virtual reality. In this preparation process, she has a new set of choices to worry about. She has a wider range of options – clothing options that may not have fit her or may have been out of her budget in the physical world. Here, the concerns that may have concerned her in the physical world – fit, comfort, cost, plausibility, are nonexistent. In this virtual party, she can look like whoever or whatever she wants to. As Annie toggles between the available choices, she realizes that her virtual self looks nothing like she does in the physical world – but does that matter?

Avatars allow for such alterations to be possible. In VR, people are visually represented by avatars – digital characters that most often have anthropomorphic features. Avatars play important roles, as they facilitate the interactions people have with others and the relationships that are fostered in these virtual spaces. Previous research has shown that avatars and how they are designed, including their nonverbal cues (e.g., gestures, postures, expressions), verbal cues (e.g., personalized language, politeness, feedback), and other visual qualities (e.g., attractiveness, anthropomorphism, rendering style, and quality) can affect the impression of trustworthiness, presence, social intimacy, and the like (Donath, 2007; Steptoe et al., 2010). Avatars play a critical role in social interaction, as being able to see what a communication partner looks like results in better performance by reducing ambiguity (Tanis & Postmes, 2007). Moreover, avatars provide different identity cues about individuals, such as their demographics and group membership.

However, in VR, avatars do not always have to match their user's actual appearance. Much like how in online spaces like the Internet an individual does not have to use their real identity (e.g., creating pseudonyms instead of using legal names), in VR one can look like anyone. Beyond modifying surface-level attributes like facial structure, hairstyle, or outfit, a person can change fundamental qualities of the self, such as genders, ethnicities, or even species. However, unlike online spaces where people might be limited to the icons or text-based representations of the self, in VR people can be represented by and embodying 3D avatars. In other words, people *become* the avatar that is representing them. This is a particularly important distinction from other digital spaces like the Internet, as now there is an added element of embodiment. Embodiment, or the psychological connection between a person and their avatar, makes a person feel as though that the body that they are in is their own.

Interestingly, it has been found that people conform their behaviors and attitudes to that of their avatars. According to a phenomenon known as the Proteus Effect (Yee & Bailenson, 2007), people tend to match their behavior to their digital self-representation, independent of how others perceive them. Across two studies, Yee and Bailenson (2007) set out to see if the attractiveness and height of an embodied avatar influenced behavior. In the first study, participants wore avatars whose faces were either attractive, unattractive, or average. After participants spent a few minutes doing exercizes in front of a virtual mirror to familiarize themselves with their new faces, they were asked to introduce themselves to a virtual confederate. During this interaction, the researchers tracked the interpersonal distance between the participant and the confederate, as well as amount of self-disclosure. Results showed that participants who wore attractive avatars walked significantly closer to the confederate and revealed more pieces of information than those in the unattractive avatar. In the second study, participants wore avatars that were either the same height as, shorter, or taller than the confederate. The participant and confederate would then play a few rounds of a money-sharing task. During these rounds, their negotiation behavior was observed. Results showed that participants who were in the taller avatar were significantly more likely to offer an unfair split than those who were shorter than or the same height as the confederate. Meanwhile, participants who wore shorter avatars were more likely to accept an unfair split than those who were shorter than or the same height as the confederate.

The Proteus Effect has also been tested in other contexts. Other researchers have found that the characteristics of your avatar can affect your dating partner choices, antisocial behavior, food choices, racial bias, financial risk-taking, and consumer choices (for a full review, see Ratan et al., 2019). In other words, *what* you look like matters, because it ultimately shapes what you believe about yourself, how you behave, and how you treat others.

Why do we change our behavior in virtual spaces? There are several theories speculating why we adjust our behavior in such ways. Some believe that it is due to an enhanced effect of self-perception (Bem, 1972), the idea that people develop attitudes by inferring them from observations of their own behavior under circumstances in which they occur. Given that virtual environments are anonymizing in nature, external cues become salient, such as avatars' appearance. People would become particularly sensitive to social cues associated with their identity that they infer from their avatar. Others have suggested that people are primed by external cues, and think and behave in ways associated with their stereotypes (Peña et al., 2009).

What are the long-term consequences of embodying avatars that do not inherently resemble you? As aforementioned, in VR users are not represented by static 2D images or icons. In VR, the user becomes their avatar. Users can see their arms, hands, torso, and depending on the platform, their legs. And, perhaps if technological advancements go in this direction, we will be able to receive feedback that gives us the sensation of touching our body's skin, hair, and its other textures. How does spending hours embodying multiple identities and adjusting your behavior in intended and unintended ways influence what you think about yourself and the world? In the same vein, given that we can easily switch between features in avatars, do we adapt our behaviors with the snap of our fingers? How do we begin to understand and measure these changes?

This experience of embodying inaccurate avatars may affect different people to varying degrees. Many social VR platforms provide limited customization options, which inevitably leads to inaccurate representations. These limited options can be seen in the lack of options for skin color, facial features, hairstyles and textures, and others. DeVeaux and colleagues (2023) touch on how there is yet to be research that investigates the unique consequences of having limited avatar racial representation in immersive environments. In interviewing users on VRChat, a popular social VR platform that allows users to upload their own avatar creations, they found that there was a dominance in light-skinned avatars and resistance in designing non-white avatars from avatar creators (e.g., due to technical challenges of designing and rendering features such as hair textures). They also reported that avatar choices were shaped by racial harassment, with several Black users hiding their race as a survival tactic. If such trends and preferences shape the avatar economy in social VR platforms, how will expectations and habits that people adopt from such environments manifest in the physical world? We have seen time and time again the issues that arize from being in spaces, communities, and institutions that are dominated by homogenous groups (e.g., people of one race, ethnicity, class, economic class, etc.). However, in VR platforms, the avatar makeup is no longer defined by features that people are inherently born with (e.g., skin color, physical attributes) or factors that cannot easily be controlled (e.g., gender, wealth), but by the designs made by the platform and the trends shaped by the users. In other words, cases of discrimination, biases, stereotypes may be more evident, prevalent, and made consciously.

Are Avatars People or Objects?

In addition to understanding how embodying avatars affects our behavior, we have to also consider our behavior *towards* other avatars. We know very well that harassment, bullying, and other harmful practices are commonplace in virtual spaces like the Internet. Much like the anonymizing nature of usernames, avatars, and their reduced cues can provide varying levels of anonymity. Avatars can empower us to say things, do things, and treat others in ways that we normally would not in the physical world.

How do avatars promote behaviors that may go against how we normally behave? Are our standards for interacting with avatars different from those we have for physical people? In other words, do we view avatars the same way we view real people? Or avatars lie somewhere in between, closer to objects than people?

Turning to classic theories within communication, Reeves and Nass' Media Equation theory proposes that we respond to media the way we respond to real people (Reeves & Nass, 1996). We respond socially and naturally to media, despite knowing that it is not reasonable to do so. We are able to form social relationships and trust media artifacts, as we perceive them as possessing human characteristics like motivation and personality (Benbasat & Wang, 2005). However, how well does the Media Equation apply to situations in which both communication partners are represented by embodied avatars? It could be argued that embodied avatars are something beyond media and are an entity that bridges the physical world with media. Similar to how we study the properties of the physical world and the unique effects of media with separate lenses, there may be parameters unique to embodied avatars that we are overlooking.

In his book on the mechanisms of moral disengagement, Bandura (2015) lays out the ways in which we convince ourselves that ethical standards do not apply to us in certain situations. Bandura details four principles of moral disengagement: reconstructing conduct, displacing responsibility, misrepresenting injurious consequences, and blaming the victim. The fourth principle asserts that we place blame onto the victim through attribution and dehumanization. This process of dehumanization includes portraying images and descriptions of people who are going to be the victims of one's actions as animals, beasts, or other subhuman objects. This act of dehumanization transforms potential empathy a perpetrator may feel. In social VR, portraying people as avatars is a literal dehumanizing process, as a person and their intricacies are being transformed into pixels with limited cues. A virtual representation of someone cannot be weighted as equally as a real person. In its absolute state, an avatar is neither tangible nor alive – it is a quasi-human being with no soul or physical body (O'Tierney, Kavanagh & Scally, 2019). If the act of representing a person as an avatar is dehumanizing, does this promote the moral engagement Bandura describes, and potentially explain why harassment is so rampant within virtual spaces (Freeman et al., 2022)?

In the same vein, in a study published in 2007, Harris and Fiske investigated how certain social groups are differentially processed in the medial prefrontal cortex (mPFC), which is activated during social cognition tasks. In this study, the authors found that there is reduced mPFC activity in response to people from social groups that elicit emotions such as disgust. This reduced mPFC activity suggests that people in certain groups are perceived differently, perhaps even less human. The takeaway from this study is that we treat people differently depending how we see and understand them. How do these results translate to avatars and who or what do they represent? It is critical that we understand if we respond to avatars in a way that dehumanizes them. Given that avatars mediate the communication between two parties, if any dehumanization does take place, we need to understand how that influences the interaction.

If it is the case that we fundamentally perceive avatars as something less than human, returning to the Media Equation theory, we are not necessarily treating avatar – the representation of people on media – the same way we would a person in the physical world. In fact, we may be treating avatars as something closer to objects. In a study conducted by Riedl and colleagues (2011), the authors investigated what trust looks like on a behavioral and neurological level when people interact with humans or avatars. In their study, they had people play an economic game in which they evaluated the trustworthiness of different humans and avatars after observing how they handled invested money. Behaviorally, people's trust in avatars was similar to that in humans. However, neurobiologically, the brain distinguishes between avatars and humans. Their results showed higher activation in a brain network that is associated with mentalizing, suggesting that people attribute the concept of a "mind" to humans, but not to the same degree to avatars.

Does this mean that we will mistreat virtual beings? In 2006, Slater and colleagues replicated Milgram's obedience study (1963) in VR. In Milgram's study, a participant is tasked to administer a memorization test to another participant in a separate room. If the second participant – a confederate, unbeknownst to the actual participant – provided an incorrect answer, the participant had to deliver an electric shock, with the voltage of the shock increasing with every wrong answer. With every shock, the confederate would cry out in pain and ask to leave the study, eventually "passing out" as the voltage got higher and higher. When the participant hesitated to continue with the study, the experimenter told the participant that they must continue. The study ended when the participant refused to obey the experimenter or when they reached the highest voltage. The results showed that the majority of the participants obeyed the experimenter and went to the final voltage. In Slater and colleague's VR replication of the Milgram study, the confederate was an agent, a computer-controlled digital representation.[2] In this study, the goal was not to look at obedience (i.e., the experimenter did not tell the participant to continue), but to see if inflicting harm on a virtual being caused anxiety, despite knowing that the virtual being and the shocks were not real. The authors recorded participants' subjective, behavioral, and physiological responses and found that people responded realistically on all three levels toward the virtual being.

There is still a ways to go with researching how we treat avatars differently from how we treat people. While there is a plethora of research on how we interact with various types of media (e.g., see the work done on Nass' Computers Are Social Actors paradigm), there remain many questions on *embodied* avatars. Especially given the critical role that avatars play in social VR settings, we must understand if we view these interactants the same way we view fellow interactants in the physical world. And if not, how our behaviors may change for better or worse.

Virtual Realism

Real, enhanced. Oftentimes people comment on how "realistic" VR is. Here, realism holds many meanings: it can refer to the photographic realism of the scene (i.e., the quality of the rendering), the anthropomorphic realism of the avatars (i.e., the human-like or non-human-like attributes of a person), or the behavioral realism of the avatars (i.e., how much the behaviors of avatars match those expected in the physical world). While there is a consistent sense of realism in the physical world, these varying degrees of realism within VR lead to different ways for us to process virtual experiences.

Consider this scene: Annie is sitting on a bench in a park. Everything around her looks oddly hyperrealistic – she can see every individual blade of grass and how it is shaded, the pores of the person sitting next to her, and the details of the leaves stuck on the windshield of a car passing by across the park. The person next to Annie is looking at her as she talks and has not broken eye contact yet. While the attention is nice, the constant eye contact from this uncanny valley-inducing person makes her shift in discomfort. Annie decides to step away, and as she is doing so, she backs into another person behind her. However, instead of colliding with the person, both Annie and the other person are standing in the same spot. Even weirder, she can see *inside* the person – she can see their teeth, tongue, and eyeballs.

This scenario is not one that is too otherworldly, as in, the people and place were recognizable as ones found in the physical world. At the same time, there were elements of this world and how it operated that were different from what was expected. These differences, some subtle, some large, ultimately lead to affect how one goes about in this alternate world.

Such scenarios are not uncommon in social VR. Many platforms value photorealistic graphics and will put users in environments with incredible details and colors that they may not notice in the physical world but are "forced" to see in the virtual world. Similarly, changes may be made to how other people are rendered, such as how they maintain eye contact or what their facial expressions look like, to enhance user experience.

However, these worlds are not perfect, and bugs may arize: malfunctions in rendering, lack of boundary setups, or badly designed features. Ultimately, these features shape the ways in which people act, think, and navigate in the virtual world.

Such transformations on behavior are unintended, as they were implemented top down by the system and cannot be controlled by the user. These features of enhanced realism, supplemented by bugs and edge cases, lead to changes in how people experience the world. Freeman and colleagues (2022), for example, describe the phenomenon of embodied harassment, a novel form of harassment in virtual social spaces, in which "harassing behaviors are both conducted and experienced through sense of embodiment about one's virtual body, such as a higher awareness of body ownership and more physical and transformative/interactive experiences" (85: 22). The authors describe how in social VR, users are not simply viewing their activities on a screen as in online gaming or traditional 2D-based virtual worlds. Users are actively engaging in the virtual space with both their virtual and physical bodies and bringing in their own history, background, and sociocultural experiences. As Blackwell and colleagues (2019), who are also researchers in the field, put it,

> Users are embodied in avatars that move when the player moves and interact with other players in three-dimensional spaces, enabling violations of personal space and corporeal presence that feel fundamentally different than interactions that occur in other online environments—an experience made potentially more salient by the unique sensation of presence, or the feeling of truly "being there".
>
> (100: 4)

A real case of such enhanced, embodied harassment was described by Zheng and colleagues. In their 2023 study, they describe an instance in which a user fell to the ground after being virtually slapped by another user, suggesting that violence in VR can trigger realistic sensations and elicit real-world responses. Such unintended affordances give rise to novel ways of existing, experiencing, and doing that may be powerful yet dangerous.

In the same vein, there are cases in which users intentionally make use of the designs of a platform to perform behaviors that are either unique to that medium or cannot be demonstrated in the physical world. In the 2023 study described earlier by Zheng and colleagues, the authors analyzed transcripts and comment sections of hundreds of YouTube videos on VR content to observe what kind of social norms form in such spaces and identified safety risks unique to social VR. Of these risks they identify, many include scenarios in which the users make intentional decisions to use VR to perform morally ill decisions. The authors list examples such as kinetic cues (e.g., poking or trampling a person's face) and proximity cues (e.g., rushing back and forth to crash into people) that users take advantage of to torment others. Likewise, people can use tools within the platform to harass others. In what the authors describe as *virtual crashing*, users use tactics such as adding particle effects like fire spawn animation to cause damage and ruin others' experiences. Given many, if not all, platforms come with some set of tools in which users can spawn new objects into the world, it is no surprize that some users will misuse these tools to terrorize others.

Another example of user-intended harassment occurs through role-plays, the act of taking on and acting out a new personality, values, and goals that are different from your own. Role-playing in social VR is easy, given users can switch between avatar appearances with a few clicks. Zheng and colleagues describe how, in VR, it is challenging to identify the seriousness and intention of the role-players. Other bystanding users may not be able to tell if the role-players are playing around or having genuinely unpleasant experiences. By allowing users to easily put on new identities and completely mask their true self, platforms not only enable the customization of identities, but they also allow users to engage behaviors that would have otherwise not been performed in the physical world.

Are VR Memories Just as Real?

So far, we've discussed actions that occur within the virtual world. A phenomenon like *virtual crashing* makes sense within the virtual context, but surely such actions do not manifest within the physical world. Why should we be concerned about events that occur inside the virtual world? This raises the question: how do lived virtual experiences change physical behavior, if at all? Beyond eliciting immediate reactions like physically reacting to a virtual punch, can the actions that we perform, witness, or experience influence how we live our lives in the physical world? Can we acquire new knowledge or behavioral changes from virtual experiences that we apply either knowingly or unknowingly into our physical lives? We've seen time and time again from previous research that virtual actions have physical world consequences (for example, prosocial behavior, Herrera & Bailenson, 2021; cultivating empathy and racial awareness, Roswell et al., 2020; climate change awareness, Fauville et al., 2020).

Research hints that VR has the power to alter even our most fundamental moral values. Segovia and colleagues (2009) put participants in VR to evaluate how viewing moral or immoral actions influenced their self-rating of morality and behavior. They showed participants scenarios of an avatar either distributing first aid kits to avatars or punching other avatars and having the bodies piling up as they fell. Those who saw the immoral behavior were more likely to take more hand sanitizer (i.e., engage in physical cleansing) and rated themselves as less moral compared to those who witnessed the moral behavior. The authors end with the following question: how does viewing immoral behavior in virtual environments affect future behavior in the physical world? Can what we see and believe in VR be powerful enough to reshape our core values on what is good or bad?

One population that may be most vulnerable to this is children. Maloney and colleagues (2020) set out to watch how minors exist and interact in social VR. They observed that children tend to perceive their interactions with other children in social VR as more realistic than those in traditional virtual worlds. They also suggest that children may have difficulty differentiating from the offline and online world, as they are still developing their personality and understanding of the self, others, and the world. Furthermore, other research has shown that it is quite easy to fabricate memories ("false memories") without the consent or mental effort of an individual. Digital media, such as those experienced in social VR, can affect our memory and emotion, and this is especially true for children, who are the most vulnerable (Segovia & Bailenson, 2009). Zheng and colleagues (2023) did find cases in which minors engaged in behaviors such as excessive cursing, which may have resulted from them imitating inappropriate behaviors displayed by adult users. In social VR, children, and adult users co-exist, meaning behaviors that adults engage in without much thought or consideration can be picked up by children and mirrored, and ultimately transferred to the physical world. Despite efforts to protect children in such spaces, it is challenging to stop them from engaging in and witnessing inappropriate behaviors.

In his book *Experience on Demand*, Bailenson (2018) touches on the effectiveness of VR in transferring skills and knowledge. The visceral nature, graphics, and interactivity of VR make it an attractive medium of training. At the same time, this could mean that other skills, such as shooting or punching, may also be transferred. One suggestion is to make VR rules operate differently than the ones in the physical world so that we don't acquire certain skills. Platform designers should consider how we can manipulate how weapons are designed and operated such that, should a VR user encounter a weapon in the physical world, they do not know how to use it. Instead of designing platforms such that users are deprived of certain behaviors, should efforts be made to change how reality functions within VR such that the way it operates is entirely unique and hard to transfer to the physical world? While this is by no means a perfect solution, it is one potential way of approaching this concern.

What Is the Technology Capable of?

These harassing phenomena and actions researchers identified are unique within social VR, and their impacts are enhanced by the affordances provided by VR. As technology continues to advance, experiences may become increasingly photorealistic, behaviorally realistic, and immersive. Technological advances in graphical rendering allow for real-time raytracing (Unity), foveated rendering (Patney et al., 2016), and simulations that retain properties in physics (Xiang et al., 2022). Beyond visual properties, there are other sensory feedback that add to a heightened sense of realism and ultimately lead to unique perceptual experiences and behaviors.

In our large-scale, longitudinal study on how avatar appearance and environmental context influence perceptions and behavior, we put hundreds of students inside social VR (Han et al., 2023). For eight weeks, students met in small groups for discussions on a platform called ENGAGE. Hosting hundreds of people inside a shared, networked environment can be challenging, as a lot of unpredictable issues, many of that are unique to VR, can arize. One procedural issue we had to grapple with was audio. Audio can be tricky to do well for several reasons. First, sound gets picked up easily. This means that whatever is going on in the background, from passing cars, phones vibrating, and conversations happening in the hallway, can seep into the virtual world. However, unlike the physical world, in which these sounds are continuous streams with varying levels of volume that depend on distance and location in space, these sounds in the virtual world are somewhat inconsistent, and picked up in pieces. Although a user has the option to turn on spatialized audio, meaning the source of a sound is localized in space, this spatialization is only preserved in the virtual world. Imagine interacting with others in a world and hearing hundreds or thousands of irrelevant sounds blasting through the headset speakers.

During the student group discussions, anyone who was not speaking stayed muted and students who had forgotten to mute themselves and had external audio seeping in were asked to turn off their microphones. Interestingly, unique behaviors arose when students wanted to indicate that they were about to unmute. DeVeaux and colleagues (2022) detailed some of these behaviors that emerged during the VR discussions:

> Other behavioral cues unrelated to real life emerged as well. Muting in ENGAGE requires pressing a button that appears on a person's wrist. Therefore, watching someone motioning their hand towards their wrist was an indicator that they were about to speak and the rest of the class would turn their heads in that student's direction.

(9)

In a non-study context, the reasons audio may be troubling are slightly different. If all audio that is picked up is broadcasted to everyone in a shared space, this means that verbal abuse, slurs, or inappropriate noizes (e.g., moans, grunts) are included. Will others use this as an opportunity to bombard other users with messages or unwanted noizes? If such is the case, how should platforms be designed such that certain sounds are picked up and prioritized? What sounds will make it to the virtual world? Will this influence how people share (or do not) sound?

Second, one of the unique affordances of VR is that it can preserve audio spatiality. When a person says something, you can tell where the sound is coming from and how far away the sound is. This enables multiple conversations to take place without having people having to remain unmuted. This also means that sounds can "sneak up" on a user. Consider this: a person comes up from behind and whispers something close to your ear. This experience may be startling or unnerving. Alternatively, you may have never experienced this. Such behaviors are not commonplace in the physical world. Going up behind people, violating their personal space, and saying something in such close proximity is not a norm. In the physical world, we have an understanding of maintaining distance between our communication partners. In the virtual world, the norms are a little different. Many social VR platforms have designed features that allow users to have a personal bubble that protects them from people coming into their personal space. Multiple people can be standing in the same exact spot, twisted inside one another. Much like how you can experience space violation, you can also experience audio violation. Sounds such as heavy and close breathing can feel visceral and uncomfortable. This feature, much like how space has been used for virtual violence (Zheng et al., 2023), can also be used to attack others.

In addition to audio, there is haptic feedback. Gloves, vests, shoes, arm sleeves, and other wearables have been developed to give users the sensation of touch. By sending electrical pulses across the skin, these wearables allow users to feel sensations ranging from punches and gunshots to caresses and hugs, ultimately creating a sense of realism and immersion. If we can punch people and know that the punches are not "real" but felt, would we do it? If we can push someone off a cliff without them facing physical damages, would we do it? If we could indirectly experience harm without the consequences, what really motivates us to not do so?

As technology continues to advance, more unique ways of knowing and experiencing are being incorporated into VR. We also see advances in other sensory feedback such as smell (see work on olfactory cues by Li & Bailenson, 2017) or taste (see work on thermal taste by Karunanayaka et al., 2018). Such technologies will introduce new ways of experiencing processes we are already familiar with from the physical world. But how do these virtual transformations transform the way we live within the virtual world?

The Future of Virtual Cultures

Taking a step back, it is easy and perhaps not fruitful to fall into panic about morality in VR, as there are still many questions that have yet to be answered in this space. What about the children? What about the violence? What about the harassment? The concern is that the metaverse, as it stands now, is under-regulated. As Blackwell and colleagues (2019) assert, because expectations for appropriateness are unclear in these environments, users and moderators are reluctant to make assumptions of people and their intentions. Furthermore, there is no certainty in what kind of user behaviors are to be expected, what kind of consequences those behaviors lead to, and who must be held accountable for those behaviors. This uncertainty stems largely from a lack of understanding of behavior in virtual environments – namely morally questionable behaviors – and what mediates them.

There are a few avenues of research that may help us understand this landscape better. First, further research is necessary on how social VR use changes with time. Curiosity and the desire to experiment with the new ways of existing in VR may drive behavior for novel users. Virtual cultures, however, develop over time. Findings concluded from a single or limited exposure to VR may not meaningfully inform how these virtual cultures will develop. Previous research has shown that time plays a critical role in painting a better picture of how we understand behaviors that take place in VR (for examples, see Bailenson & Yee, 2006; Han et al., 2023). To truly understand the cultures within the metaverse we not only need experiments that collect responses at multiple different time points, but also ethnography that investigates already-existing cultures and how their members, values, and rules have come to be.

Second, we must grapple with the question of *why*. Why do our behaviors change within the metaverse? Are we acting out roles that we have come to learn from the physical world (i.e., the Proteus Effect)? Or are we satisfying a curiosity that would otherwise be impossible in the physical world? Or are we responding to a new set of rules laid out by the environment and the medium? Or are we acting out of courage that anonymity and the ability to change between multiple selves provide within the metaverse?

Third, we need to think about what happens outside of VR. By this, we don't necessarily mean how our behaviors are affected by our virtual experiences. Rather, what are we doing with this new information we have in our virtual lives? We've now seen that the behaviors enacted, and ultimately the cultures that form, within VR are different. What can we make use of that information? Perhaps one of the biggest differences between VR and the physical world is that the former is entirely created, controlled, and surveilled by programmers, developers, designers, and the other. If surveillance and what we do with people's data is a concern in the physical world, it should be an even bigger concern within VR. Whereas in the physical world, there may be blind spots where a camera isn't present, or a sensor isn't working, or a tracker isn't available yet, in VR, all our movements, physiological responses, and decisions can be easily tracked. Studies have shown that we can identify people with pretty high accuracy by simply looking at how they move (Miller et al., 2020). Beyond movement, there is a lot of biometric data, from eye-gaze, pupillometry, heart rate, and facial muscle movement, that can be collected from these new, advanced headsets that are being continuously released. In the metaverse, where we can become anyone or take on multiple identities (i.e., have multiple avatars), how should a person's profile – or multiple profiles – be created and understood?

Notes

1 Throughout this chapter the terms social VR and the metaverse interchangeably. Here, both terms refer to a shared social virtual environment accessed through a VR headset in which people interact in real-time and are represented by avatars.
2 An agent is different from an avatar such that an avatar is a digital representation controlled by a person.

References

Bailenson, J. (2018). *Experience on Demand: What Virtual Reality Is, How it Works, and What it Can Do*. W. W. Norton & Company

Bailenson, & Yee. (2006). A longitudinal study of task performance, head movements, subjective report, simulator sickness, and transformed social interaction in collaborative virtual environments. *Presence: Teleoperators and Virtual Environments*, 15(6), 699–716. https://doi.org/10.1162/pres.15.6.699

Bandura, A. (2015). *Moral Disengagement*. Worth Publishers

Bem, D. J. (1972). Self-perception theory. In L. Berkowtiz (Ed.), *Advances in Experimental Social Psychology* (Volume 6, pp. 1–62). Elsevier. http://dx.doi.org/10.1016/s0065-2601(08)60024-6

Benbasat, I., & Wang, W. (2005). Trust in and adoption of online recommendation agents. *Journal of the Association for Information Systems*, 6(3), 4

Blackwell, L., Ellison, N., Elliott-Deflo, N., & Schwartz, R. (2019). Harassment in social virtual reality. *Proceedings of the ACM on Human-Computer Interaction*, 3(CSCW), 1–25. https://doi.org/10.1145/3359202

DeVeaux, C., Han, E., & Bailenson, J. N. (2022). Expanding education through virtual reality. In S. P. McKenzie, L. Arulkadacham, J. Chung & Z. Aziz (Eds.), *The Future of Online Education* (pp. 325–336). Nova Science Publishers

DeVeaux, C., Han, E., Landay, J. A., & Bailenson, J. N. (2023, May). A presence of absence: Understanding disparities in avatar racial representation and embodiment in social VR. 73rd Annual International Communication Association Conference.

Donath, J. (2007). Virtually trustworthy. *Science*, 317(5834), 53–54. https://doi.org/10.1126/science.1142770

Fauville, G., Queiroz, A. C. M., & Bailenson, J. N. (2020). Virtual reality as a promising tool to promote climate change awareness. In J. Kim & H. Song (Eds.), *Technology and Health* (pp. 91–108). Elsevier. http://dx.doi.org/10.1016/b978-0-12-816958-2.00005-8

Freeman, G., Zamanifard, S., Maloney, D., & Acena, D. (2022). Disturbing the Peace: Experiencing and mitigating emerging harassment in social virtual reality. *Proceedings of the ACM on Human-Computer Interaction*, 6(CSCW1), 1–30. https://doi.org/10.1145/3512932

Han, E., Miller, M. R., DeVeaux, C., Jun, H., Nowak, K. L., Hancock, J. T., Ram, N., & Bailenson, J. N. (2023). People, places, and time: A large-scale, longitudinal study of transformed avatars and environmental context in group interaction in the metaverse. *Journal of Computer-Mediated Communication*, 28(2). https://doi.org/10.1093/jcmc/zmac031

Harris, L. T., & Fiske, S. T. (2007). Social groups that elicit disgust are differentially processed in mPFC. *Social Cognitive and Affective Neuroscience*, 2(1), 45–51. https://doi.org/10.1093/scan/nsl037

Herrera, F., & Bailenson, J. N. (2021). Virtual reality perspective-taking at scale: Effect of avatar representation, choice, and head movement on prosocial behaviors. *New Media & Society*, 23(8), 2189–2209. https://doi.org/10.1177/1461444821993121

Karunanayaka, K., Johari, N., Hariri, S., Camelia, H., Bielawski, K. S., & Cheok, A. D. (2018). New thermal taste actuation technology for future multisensory virtual reality and internet. *IEEE Transactions on Visualization and Computer Graphics*, 24(4), 1496–1505. https://doi.org/10.1109/tvcg.2018.2794073

Li, B. J., & Bailenson, J. N. (2017). Exploring the influence of haptic and olfactory cues of a virtual donut on satiation and eating behavior. *Presence: Teleoperators and Virtual Environments*, 26(3), 337–354. https://doi.org/10.1162/pres_a_00300

Maloney, D., Freeman, G., & Robb, A. (2020, November 2). A virtual space for all: Exploring children's experience in social virtual reality. *Proceedings of the Annual Symposium on Computer-Human Interaction in Play*, 472–483. http://dx.doi.org/10.1145/3410404.3414268

Milgram, S. (1963). Behavioral study of obedience. *The Journal of Abnormal and Social Psychology*, 67(4), 371–378. https://doi.org/10.1037/h0040525

Miller, M. R., Herrera, F., Jun, H., Landay, J. A., & Bailenson, J. N. (2020). Personal identifiability of user tracking data during observation of 360-degree VR video. *Scientific Reports*, 10(1), 17404. https://doi.org/10.1038/s41598-020-74486-y

O'Tierney, A. J., Kavanagh, D., & Scally, K. (2019). Me and my avatar: Acquiring actorial identity. In H. Hwang, J. A. Colyvas & G. S. Drori (Eds.), *Agents, Actors, Actorhood: Institutional Perspectives on the Nature of Agency, Action, and Authority* (pp. 65–86). Emerald Publishing Limited. http://dx.doi.org/10.1108/s0733-558x20190000058006

Patney, A., Salvi, M., Kim, J., Kaplanyan, A., Wyman, C., Benty, N., Luebke, D., & Lefohn, A. (2016). Towards foveated rendering for gaze-tracked virtual reality. *ACM Transactions on Graphics*, 35(6), 1–12. https://doi.org/10.1145/2980179.2980246

Peña, J., Hancock, J. T., & Merola, N. A. (2009). The priming effects of avatars in virtual settings. *Communication Research*, 36(6), 838–856 https://doi.org/10.1177/0093650209346802

Ratan, R., Beyea, D., Li, B. J., & Graciano, L. (2019). Avatar characteristics induce users' behavioral conformity with small-to-medium effect sizes: A meta-analysis of the proteus effect. *Media Psychology*, 23(5), 651–675 https://doi.org/10.1080/15213269.2019.1623698

Reeves, B., & Nass, C. (1996). *The Media Equation: How People Treat Computers, Television, and New Media Like Real People.* CSLI Publications.

Riedl, R., Mohr, P., Kenning, P., Davis, F., & Heekeren, H. (2011). Trusting humans and avatars: Behavioral and neural evidence. 32nd International Conference on Information Systems

Roswell, R. O., Cogburn, C. D., Tocco, J., Martinez, J., Bangeranye, C., Bailenson, J. N., Wright, M., Mieres, J. H., & Smith, L. (2020). Cultivating empathy through virtual reality: Advancing conversations about racism, inequity, and climate in medicine. *Academic Medicine*, 95(12), 1882–1886. https://doi.org/10.1097/acm.0000000000003615

Segovia, K. Y., & Bailenson, J. N. (2009). Virtually true: Children's acquisition of false memories in virtual reality. *Media Psychology*, 12(4), 371–393. https://doi.org/10.1080/15213260903287267

Segovia, K. Y., Bailenson, J. N., & Monin, B. (2009). Morality in tele-immersive environments. Proceedings of the International Conference on Immersive Telecommunications (IMMERSCOM), May 27 – 29, Berkeley, CA, USA. http://dx.doi.org/10.4108/icst.immerscom2009.6574

Slater, M., Antley, A., Davison, A., Swapp, D., Guger, C., Barker, C., Pistrang, N., & Sanchez-Vives, M. V. (2006). A virtual reprise of the Stanley Milgram obedience experiments. *PLoS ONE*, 1(1), e39. https://doi.org/10.1371/journal.pone.0000039

Steptoe, W., Steed, A., Rovira, A., & Rae, J. (2010, April 10). Lie tracking. Proceedings of the SIGCHI Conference on Human Factors in Computing Systems. http://dx.doi.org/10.1145/1753326.1753481

Tanis, M., & Postmes, T. (2007). Two faces of anonymity: Paradoxical effects of cues to identity in CMC. *Computers in Human Behavior*, 23(2), 955–970. https://doi.org/10.1016/j.chb.2005.08.004

Xiang, D., Bagautdinov, T., Stuyck, T., Prada, F., Romero, J., Xu, W., Saito, S., Guo, J., Smith, B., Shiratori, T., Sheikh, Y., Hodgins, J., & Wu, C. (2022).

Dressing avatars. *ACM Transactions on Graphics*, 41(6), 1–15. https://doi.org/10.1145/3550454.3555456

Yee, N., & Bailenson, J. (2007). The proteus effect: The effect of transformed self-representation on behavior. *Human Communication Research*, 33(3), 271–290. https://doi.org/10.1111/j.1468-2958.2007.00299.x

Zheng, Q., Xu, S., Wang, L., Tang, Y., Salvi, R. C., Freeman, G., & Huang, Y. (2023). Understanding safety risks and safety design in social VR environments. *Computer Supported Cooperative Work*, 7(CSCW1), 1–37. https://doi.org/10.1145/3579630

10 Doing Good With Virtual Reality

The Ethics of using Virtual Simulations for Improving Human Morality

Jon Rueda

Introduction

Can Virtual Reality (VR) become a force for good?[1] The possibility that virtual experiences can lead to positive social change in the non-virtual world is exciting. VR comprises a set of technologies that enable vivid, interactive, and immersive experiences in digitally created environments (Slater & Sanchez-Vives, 2016; Bailenson, 2018).[2] If experiences in virtual environments could induce beneficial attitudinal and behavioral changes in the "real physical world", these technologies could become important allies in achieving significant socio-moral goals. Still, assuming this promising possibility could mean accepting its less pleasant counterpart—that VR technologies can promote bad moral traits and lead to ethically problematic dispositions in users.

In this chapter, I philosophically address the impact of VR on human morality. This issue is crucial from an ethical perspective in part because the (un)desirability of many virtual practices must be judged in terms of their benevolent or malevolent influence on the development of users' moral character and conduct (Cotton, 2021). Indeed, the consequences of behavioral changes are particularly relevant with this technology. In contrast with other more passive media, the fact that in VR the participant is not a "spectator" but an "actor" raises the possibility that virtual agency could have consequences on real-world behavior (Brey, 1999, p. 8). Furthermore, the considerable cheapening and diffusion of VR technologies (including rudimentary cellphone-based hardware formats), with the consequent boom in domestic use, make this issue even more pressing. Virtual experiences—seemingly confined to the realm of the home—are becoming massive, so these presumably private practices could have ethically relevant public consequences.

In assessing the potential for VR to do good, a number of empirical and normative issues need to be considered. This is precisely the purpose of the overview I propose in this chapter. On the one hand, if we want to know

DOI: 10.4324/9781003359494-14

what hopes and fears are justified, we must attend to the burgeoning inter-disciplinary literature studying the impacts of VR on people's prosocial be-haviors and moral capacities. Knowing which types of virtual experiences produce changes in attitudes, which reduce or increase implicit biases, or which factors of virtual exposures may cause lasting effects is a straight-forward empirical question. On the other hand, whether these changes are desirable or undesirable is a normative problem. Even if there is agree-ment on the evidence generated on whether VR could have a particular influence, there may still be ethical divergences on whether the influence in question is good or bad. These empirical and normative levels will be intertwined in the subsequent discussion, showing that the assessment of the impact of VR on human morality is complex and nuanced.

The structure of this chapter is as follows. I will begin by showing the rationale for using VR to improve human morality, pointing out the spe-cific potential of this technology. Next, I will look at three cases where VR could notably influence morally significant abilities and prosocial types of behavior. I will then show that socio-moral projects through VR also have objectionable aspects by analyzing three specific problems. Finally, I will conclude with a series of reflections on the limitations of the chapter and with future lines of discussion.

The Rationale for Using VR to Improve Morality

Human morality can be empirically studied and deliberately influenced through emerging technologies. On the one hand, knowledge about the foundations of morality in our species has considerably increased in re-cent decades. First, human morality has a neurobiological basis (Rueda, 2021). We know, for example, that certain brain lesions—either caused by accidents or by non-traumatic disorders—can affect our socio-moral capacities (Harlow, 1848; Damasio, 1994; Mendez, 2009). Neuroimaging studies also provide valuable information about the brain in action, show-ing which brain regions are more important in specific moral judgments or in morally problematic situations (Greene et al., 2001; Illes, 2003; Ra-cine et al., 2005; Prehn & Heekeren, 2014). There is also ample evidence on the neurobiological basis of morality because it has been shown that neurochemical modulation and brain stimulation techniques can influence moral judgments and behaviors (Levy et al., 2014; Bourzac, 2016; Crock-ett, 2016; Di Nuzzo et al., 2018). Second, other branches have studied the evolutionary origin of morality. The building blocks of our moral psy-chology seem to have developed in the Pleistocene, when humans lived in small, interdependent, and close-knit communities (van Schaik et al., 2014; Burkart, 2018; Tomasello, 2018). The moral dispositions that re-mained ingrained in our forged evolutionary psychology hinder, according to some, the fulfillment of today's ethical aspirations to care for strangers who are distant in space and time (Persson & Savulescu, 2012). Third, we have further empirical evidence coming from "experimental ethics", that is, from studies (using methods mainly from cognitive sciences, empirical moral psychology, or behavioral economics) showing the psychological, cognitive, and behavioral mechanisms that influence our moral tendencies in both lay and expert persons (see Dworazik & Rusch, 2014; Aguiar et al., 2020).

On the other hand, emerging technologies can deliberately influence, for better or worse, our moral capacities. In particular, the field of study on "moral enhancement" has investigated how to use technological advances in beneficial ways to improve human morality. By "moral enhancement", we generally mean using technological applications or science-based interventions to improve cognitive, emotional, motivational, and behavioral aspects related to morality (DeGrazia, 2014; Raus et al., 2014; Rueda, 2020). Different methods could serve these purposes, such as, among others, emerging neurotechnologies (Earp et al., 2018; Di Nuzzo et al., 2018), various pharmaceuticals (Crockett, 2014; Levy et al., 2014; Lara, 2017), genetic technologies (Faust, 2008; Walker, 2009; Agar, 2010; Douglas & Devolder, 2013), artificial intelligence (Savulescu & Maslen, 2015; Klincewicz, 2016; Giubilini & Savulescu, 2018; Lara & Deckers, 2020), and even VR itself (Rueda & Lara, 2020; Lara & Rueda, 2021).

In this chapter, I have an interest that goes beyond the use of VR for "moral enhancement". That term is contentious on both the "enhancement" and "moral" sides. To begin with, the term "enhancement" is not without controversy (Gyngell & Selgelid, 2016). For instance, a predominant conception of "enhancement" points to interventions that improve capabilities beyond what is normal (in a statistical sense) in a population (Daniels, 2000; Schwartz, 2005). In other words, an enhancement goes beyond what is typical in terms of concrete functionality in a given group. However, establishing what is "normal" or "typical" is often difficult (Gyngell & Selgelid, 2016; Rueda et al., 2021). We also risk confusing what is normal with what is desirable. Furthermore, going beyond the normal need may not always be a positive thing—increasing, for instance, certain abilities such as smell and hearing may be undesirable in many environments (Eerp et al., 2014).

It is important, moreover, to reflect on what we mean by "moral" in *moral* enhancement. It is difficult (or perhaps impossible) to offer a completely neutral (non-value-laden) description of the capabilities we regard as "moral". That is, it is challenging to articulate a definition of moral enhancement that does not implicitly have some normative positioning about what is desirable or undesirable in order to facilitate morality (DeGrazia, 2014; Raus et al., 2014). Similarly, there are many rival conceptions of what is morally valuable in our societies. Some people may think that moral enhancements would help improve compassion, while others would think of interventions that would strengthen loyalty to their nation or devotion to their family (O'Neill et al., 2022). Given these difficulties, I will not restrict the discussion below to narrow views of morality or solely to strict uses of "enhancement" in the extended (but controversial) sense mentioned above. Rather, I am interested in the varied impacts of VR on human morality in a broad sense.

Let us now consider the particularities of VR concerning its potential to affect the cognitive underpinnings of human morality. Although it has not been given the most systematic attention, VR is a very powerful technology for influencing the psychological and behavioral aspects of morality for several reasons. Immersive experiences in virtual environments can affect the plasticity of the human mind and modulate the contextual factors that shape personal conduct, even without being aware of the influence of these external factors (Madary & Metzinger, 2016). This is largely due to the feeling of "presence", a main psychological characteristic of VR, by which we have the experience of "being there" (Heater, 1992). In other words, the place illusion is crucial for having the sense of being in the virtual world (Slater, 2009). When the feeling of presence is properly attained, users may react in virtual environments in a manner analogous to how they would react in non-virtual ones (Slater et al., 2006; Felnhofer et al., 2015; Oh & Bailenson, 2017; Bailenson, 2018).

In addition to the immersive nature of virtual scenarios, it is worth mentioning the phenomenon of body ownership illusion. Head-mounted displays (HMDs) enable the simulation of virtual embodiment, namely, having the illusion that you are incorporated in the body of an avatar in the first-person perspective. Interestingly, changing the virtual body changes the self, even to the point of having after-effects in subsequent real-life experiences (Maister et al., 2015; Slater et al., 2020). Indeed, the "proteus effect" refers to the phenomenon of how altering our digital self-representation influences behavior, acting according to what may be expected from the identity of the virtual avatar—and whose effect sometimes extends beyond the virtual experience (Yee & Bailenson, 2007). For instance, being embodied in a superhero avatar when people are in danger increases helping behavior toward them (Rosenberg et al., 2013).

Another interesting feature of VR is its potential to produce an enormous range of possible experiences. Jeremy Bailenson labeled, in this sense, VR as an "experience generator" (Bailenson, 2018)—which reminds us of Robert Nozick's (1974) idea of the "experience machine". Others have spoken of VR as a powerful simulation medium (Ramirez & LaBarge, 2018). Both conceptions converge in the view that the fundamental element of VR is the type of experiences it allows us to live. We could have, albeit costly, many of these experiences in the non-virtual physical world. Others, in contrast, would be directly impossible. For example, to return to the previous mention, VR can (rudimentarily) simulate abilities such as the superpower of flight so that we can have the experience of temporarily being a superhero. In view of its experience simulator character, it is not surprising that VR is increasingly associated with a kind of "experiential ethics" (Cotton, 2021, p. 31). A prominent appeal of VR for moral improvement is therefore its potential for "experiential moral learning". VR allows us to reinforce moral qualities through doing and establishing habits, which is fundamental to moral development according to ancient Greek thinkers and contemporary pragmatists (Rueda, 2022; Rueda & Dore-Horgan, 2022). This possibility is especially seductive for subjects in limited social environments and who cannot develop actions that could improve their moral character, such as prisoners (Ligthart et al., 2022).

In light of this, VR can influence our morality by placing us in different environments that we feel are real, through body transfer to other avatars, and by providing us with morally relevant experiences. The following section will address some more concrete impacts of this technology.

Domains of Socio-moral Improvement Through VR

VR can influence human morality in many ways. In this section, I will focus on just a few domains. The following cases have been selected either because they have received considerable academic attention, because they have generated particular enthusiasm in civil society, or because they represent a mixture of both. I shall particularly focus on how VR technologies may affect empathic abilities, reduce implicit biases, and improve pro-environmental awareness and behavior.

The "Empathy Machine"

The use of VR with the intention of improving empathic skills has been one of the leading discussions. This issue is the subject of at least three fundamental controversies. The first is what we mean by empathy and what subtypes can be distinguished within it. The second is whether VR actually influences empathy. The third is if the role played by empathy, or any of its subtypes, is important for morality.

Let us start with the conceptual issue. The term "empathy" comprises a set of diverse psychological capacities that enable us to feel or imagine the experiences of others. Although at least eight subtypes can be distinguished (Batson, 2009), it is common, in the literature about VR, to focus on two very different kinds of empathy (Fisher, 2017; Ramirez, 2017; Bailenson, 2018; Francis et al., 2018; Hamilton-Giachritsis et al., 2018; Seinfeld et al.,

2018; van Loon et al., 2018; Schoeller et al., 2019). On the one hand, *emotional empathy* usually refers to the action of mirroring other people's emotions, mimicking their affective states almost as a reflex action. On the other hand, *cognitive empathy* refers to taking the other person's perspective, either by imagining the mental state of that subject from their point of view (imagine-other) or by imagining ourselves in that person's position (imagine-self) (Davis, 1980, 1983; Batson, 1997). To these two main subtypes, some also add a third *motivational* dimension to empathy (Zaki, 2014; Bailenson, 2018). Feeling or imagining the affective states of others may lead to alleviate their perceived suffering or to be proactively concerned about their well-being (Batson, 2015). This subtype, linked to altruistic motivation, is sometimes referred to as *empathic concern*.

After seeing the different ways of understanding empathy, let us now turn to the empirical question of how VR can affect these empathic skills. This is an important question because several studies have examined whether VR really works as an "empathy machine". The term "empathy machine" was used by the filmmaker and visual storyteller Chris Milk (2015) to characterize VR during a TED talk, becoming a popular expression beyond academic circles. Some organizations, visual artists, and communicators with social concerns adopted this idea of using VR to empathize with, for example, refugees (Milk, 2015) or non-human animals.[3] But were these social projects effective in influencing empathy? To answer this question, I must first clarify the main technical characteristics of these experiences.

Most of the creators from civil society used videos recorded in 360° to be projected on the HMD. These cinematic experiences may give the impression of seeing the life of their protagonists in first person, but they do not allow us to embody interactive virtual avatars as such. Evidence in hand, these immersive videos or movies in VR glasses are not as powerful in eliciting empathy as their creators would have wished, nor significantly more advantageous in influencing empathic processes than other non-immersive media (Sundar et al., 2017; Archer & Finger, 2018; Bang & Yildirim, 2018; Weinel et al., 2018). The research of Schutte and Stilinović (2017) is one of the exceptions which showed a significant increase in empathic perspective-taking and empathic concern, in which participants did empathize more with the refugee protagonist of Milk's *Clouds Over Sidra* compared to control groups viewing it in a two-dimensional format. But, on balance, the review of the majority of evidence does not allow us to affirm that immersive 360° videos make VR a particularly effective empathy machine (Rueda & Lara, 2020; Sora-Domenjó, 2022).

Other lines of research are related to virtual embodiment. In these simulations, the aim is to increase empathy with the groups represented by the avatars in which the participants were incorporated. In general terms, this line has been more successful in eliciting prosocial motivation and more inclusive attitudinal changes than immersive video. As mentioned, virtual embodiment permits us to incorporate individuals in avatars that represent the social targets with whom we want to raise empathy (Rueda & Lara, 2020). In this way, changing our body depictions and social identities in VR may increase the tendency to take the perspective of the represented collective outside virtual environments. This strategy has been used to embody participants in avatars of other genders (Seinfeld et al., 2018), skin tone (Groom et al., 2009; Peck et al., 2013; Banakou et al., 2016; Hasler et al., 2017), ages (Oh et al., 2016; Hamilton-Giachritsis et al., 2018), members of disabled groups (Ahn et al., 2013; Chowdhury et al., 2019), homeless people (Herrera et al., 2018), and even non-animal species (Ahn et al., 2016).[4]

We should be cautious, however, in interpreting too optimistically the later results as clear examples of empathy enhancement. Here, unlike with 360° immersive videos, we are not discussing whether there have been significant influences of virtual embodiments, but rather what the causal mechanisms inducing those prosocial changes are. The theoretical disagreement is about whether the main effects are produced by mechanisms linked to empathy or not. Some studies have associated positive attitudinal changes after virtual embodiment with empathy, as in one study that showed how cognitive empathy was increased (van Loon et al., 2018), and, more importantly, in another longitudinal study in which impacts on perspective-taking ability were longer lasting (Herrera et al., 2018). Others have offered different cognitive explanations. At least regarding positive changes in implicit attitudes, some views challenge the idea that empathy-related mechanisms are the underlying force of these effects (Tsakiris, 2017; Bedder et al., 2019; Slater & Banakou, 2021). I shall attend to them in the following subsection when addressing the reduction of implicit biases.

Finally, I shall briefly consider the normative dimension of the controversy around VR and empathy enhancement. Is it desirable to improve empathy to behave more morally? That question has sparked heated debates. On the one hand, for some authors, the role of empathy in morality is contested (Prinz, 2011a, 2011b; Bloom, 2016). On these views, not only is empathy not necessary for morality, but it sometimes hinders our aspirations for impartiality, as our empathic tendencies favor in-groups, i.e., those to whom we are already most motivated to help. Admittedly, we tend to have less problems in sharing the emotions of individuals with whom we share similarities, affinities, or familiarity (Bertrand et al., 2018). Therefore, according to this position, empathy may be counterproductive if we want to improve moral behavior in favor of out-groups.

On the other hand, empathy may play a positive contribution to moral behavior. This is the case of reflective and reason-guided conceptions of empathy (Persson & Savulescu, 2018; Rueda & Lara, 2020;

Lara & Rueda, 2021), which claim that empathic perspective taking is sometimes a source of valuable moral reasons to care about the well-being of others. These views, however, may serve to criticize certain proposals that use VR for empathic purposes. For instance, emotive videos in an immersive 360° format are not ethically promising as long as they do not mainly foster perspective-taking capabilities (Rueda & Lara, 2020), which are less morally controversial compared to rudimentary forms of emotional contagion (see Prinz, 2011a, 2011b; Masto, 2015; Bloom, 2016). But these views, by contrast, could positively assess virtual embodiment initiatives as far as they make it easier to engage with the perspective of those with whom they have more difficulties, i.e., with the out-groups.[5] In this way, improving empathy can be beneficial (but of course not sufficient) for morality because of its epistemological functions (Coplan, 2011; Oxley, 2011; Masto, 2015) and its role in motivating behavior (Masto, 2015; Persson & Savulescu, 2018; Read, 2019).

To summarize, VR puts users in the shoes of individuals representing multiple collectives. As the evidence and normative positions are not entirely conclusive, the debate on the use of VR to influence empathy is likely to continue.

Reducing Implicit Biases

As shown in the case of empathy, VR may alter social cognition. In this subsection, I will shortly deal with the phenomena—somewhat related to the previous controversy—of implicit biases. Generally speaking, implicit biases are automatic stereotypes and unintentional prejudices that may influence our decisions and behavior, leading to harmful social injustices (Brownstein, 2019). So, an apparently attractive option to influence our morality would be to reduce unconscious biases and implicit attitudes related to negative social stereotypes.

The process of virtual embodiment, again, is a promising way for achieving the reduction of implicit biases. By changing our bodily self-representation, VR can help decrease implicit biases regarding the collectives represented by the avatar in which users are incorporated. Tanvir I. Chowdhury and colleagues showed, for instance, that embodying an avatar in a wheelchair reduces the negative association toward disabled individuals and leads to a lower score on the implicit association test (Chowdhury et al., 2019). They also revealed that the effects of simulating disabilities were significantly greater in VR compared to a non-immersive computer desktop.

Furthermore, racial embodiment has been another prominent case of study. Tabitha C. Peck and colleagues showed that embodying light-skinned participants in dark-skinned avatars reduced their implicit racial bias. Domna Banakou and coauthors showed, moreover, that the reduction of implicit racial bias may last even one week after only one virtual exposure (Banakou et al., 2016). Similarly, a study conducted by Béatrice S. Hasler et al. (2017) demonstrated that racial embodiment may also reverse in-group mimicry favoritism, independently of their level of implicit racial bias. It increases mimicry behavior with the virtual counterpart that shares the virtual race of the participants (but not their actual racial identity). However, there is one example of increasing implicit biases after racial embodiment. In fact, in the first experiment on full-body racial embodiment, conducted by Victoria Groom and colleagues, Caucasian participants increased their implicit racial biases after being embodied in dark-skinned virtual avatars (Groom et al., 2009). Subsequent research on racial embodiment has pointed out that a possible explanation for the above result is that Groom et al. placed participants in job interviews—arguably socially hostile, competitive scenarios in which self-representation is more fragile (Hasler et al., 2017; Bedder et al., 2019; Slater & Banakou, 2021).

That said, how are the changes in implicit biases induced by VR? Implicit biases, needless to say, have a complex nature with cognitive, social, and physical ingredients (Brownstein, 2019; Lin et al., 2020). Still, some explanations from the cognitive sciences are particularly appealing to try to make sense of these changes produced by virtual embodiment. Prominently, Rachel L. Bedder and colleagues proposed the mechanistic account of "bodily resonance" to explain previous cases of implicit bias reduction through VR (Bedder et al., 2019). This proposal is based on how we compared our self-image representations with other individuals. Our self-image representations include diverse characteristics such as group membership, physical, bodily, or aesthetic traits. Bodily resonance, then, refers to the cognitive mechanism of comparing those encoded features with other people to see which characteristics overlap and which diverge. This cognitive-based explanation converges with previous contributions about how changing our bodily representations may affect our self-image and subsequent social cognition of out-groups (Maister et al., 2015; Farmer & Maister, 2017; Tsakiris, 2017). Virtual embodiment fosters, therefore, new associations between our self-representation and others, which may lead to a more positive appraisal of features that were not previously encoded in our self-image.

Let us now turn to the ethical terrain. At first glance, the possibility of reducing implicit biases seems ethically desirable. These implicit attitudes condition our decisions and conduct, even if we are not aware of them, and even if we would well intentionally wish that they did not affect us. Therefore, many agents may wish to reduce implicit biases that reinforce problematic behaviors and stereotypes. We could consider these interventions as improvements in moral autonomy.

To this positive reading, however, two cautions should be added. First, as we have seen, the context and interactions that occur in virtual embodiment seem relevant in modulating the effects. Therefore, designers should avoid generating virtual experiences that produce a reactive self-identification that increases implicit biases toward disadvantaged groups. Indeed, avoiding undesired effects is important to avoid a backlash against the socio-moral uses of VR (Sora-Domenjó, 2022). Second, although VR can serve as a bias-reduction technology to help us mitigate problematic tendencies to which we are subtly predisposed, the use of this technology to eradicate social discrimination is limited. The case of racial embodiment is a clear example in this regard. Certainly, if we are predisposed to implicit racial biases, reducing them is a good thing. It even appears, according to neuroscientific research, that racial aversion has neural correlates—being particularly related to the activity of the amygdala (Gazzaniga, 2005; Douglas, 2008, 2013). However, implicit racial attitudes should not be simply confused with the phenomenon of racism. Racism is a complex social phenomenon with structural causes, often including conscious discrimination, and where noticeable power relations are at play. Therefore, strategies to combat racism—in addition to using initiatives such as racial embodiment to reduce implicit bias—should be complemented with further measures that tackle the social roots of the problem (Rueda & Lara, 2020).

Improving Pro-environmental Behavior

Environmental degradation and climate change are among the most pressing challenges of this century. Those large-scale problems are global in scope and have repercussions for future generations. To mitigate their most deleterious ravages, these ecological challenges require joint coordination along with reinforcing individual motivation (Rueda, 2020). The massive scale of environmental problems, unfortunately, can demotivate many individuals due to the limitations of our moral psychology that favors those close in space and time (Persson & Savulescu, 2012). The difficulties in switching behavior to tackle climate change, however, are not only a motivational problem but they may also respond to cognitive factors (Kulawska y Hauskeller, 2018, p. 377). Fortunately, as we shall see, VR technologies can help in raising awareness and encouraging pro-environmental behavior.

Mainly, VR leads to the transformation of current information-based environmental communication strategies into experience-based approaches (Plechatá et al., preprint). Multiple studies have exploited the idea of using vivid virtual experiences to test the impacts of VR on pro-environmental behavior. For this purpose, influencing the *locus of control* of individuals is one promising way of increasing responsibility awareness for sustainable behavior. Environmental locus of control refers to having the internal perception that one's own behavior has a direct impact on the environment (Cleveland et al., 2005). In a seminal article, Sun Joo (Grace) Ahn and colleagues showed that the virtual experience of cutting a tree augmented the self-reported environmental locus of control and diminished paper consumption in the real world compared to print and video messages (Ahn et al., 2014).

There are more studies providing evidence on how showing the environmental impacts of personal decisions in VR may achieve behavioral changes in the non-virtual world. Participants receiving vivid messages about the energy used to heat and transport water during a virtual shower used cooler water during real-life hand washing than participants that were exposed to less vivid messages (Bailey et al., 2015). Furthermore, other strategies focus on narrowing the temporal perception of future impacts. Indeed, another way to influence environmental attitudes is accelerating the progress of time to see, for instance, how marine life is endangered by ocean acidification, which positively affects its connectedness with nature (Ahn et al., 2016). VR serves to provide, moreover, experiences intended to foster sustainable eating through the anticipation of the future bad impacts of our dietary carbon footprint. Adéla Plechatá and colleagues showed that VR was effective in changing to more environmentally friendly plant-based diets one week after the virtual exposure (Plechatá et al., 2022; Plechatá et al., preprint).

In short, influencing locus of control shows how one can affect the perception of one's own responsibility. Although moral responsibility is not exhausted in attributability, attributability is important for knowing how to delineate when an agent causally contributes to a phenomenon for which they may deserve praise or blame (Douglas, 2019). Thus, VR can narrow the attribution gap between an action and its environmental impact. This attribution gap can also be narrowed, moreover, by showing the future impacts of present actions. In this way, diverse phenomena of environmental degradation occurring in slow motion—such as marine acidification—can be perceived as events of greater proximity through its acceleration in VR. Therefore, the potential for promoting environmental conservation is not negligible (Millar, 2016).

So, if combating environmental challenges is normatively desirable, and given the beneficial impacts empirically shown in the above studies, VR can be an ethically valuable tool to improve our pro-environmental behaviors.

Objections

Initiatives that seek to do good through VR are not exempt from criticism. In this section, I will elaborate on three risks of VR, showing how (seemingly desirable) socio-moral projects can contentiously feed the complacency of the privileged, how some virtual experiences might develop undesirable moral traits or behaviors, and how there may be trade-offs for the well-being of users. While these objections have appeal, my purpose is not to argue that they are strong enough to disqualify all attempts to improve human morality through VR. Rather, what I intend to show is that these counterpoints may reduce part of the enthusiasm for using VR to improve imperfect human morality.

Swelling the Complacency of the Privileged

Doing good can make us feel good. Personal gratification is often an element of altruistic behavior—which may even lead us to ask whether there is such a thing as a totally disinterested altruistic action (May, 2011; Kraut, 2020). Long-standing philosophical controversies aside, this question is important in the case of VR projects that want to promote social good, with some particularities. Most VR users and content creators belong to high-income countries, which may give rise to some criticism.

Lisa Nakamura (2020) has particularly been very critical of projects that try to do social good originating from the VR industry. With video games and porn being the most popular applications of VR, marketing VR as a technology for empathy and justice can be a social washing strategy. According to Nakamura, several leaders of Big Tech and digital platform capitalism—including Mark Zuckerberg—have been keen to promote VR as a technology that fosters greater social connections and progress. Many immersive media practitioners and tech entrepreneurs have taken advantage of this "cultural alibi" to develop projects that offer virtual experiences from the perspective of people from disadvantaged groups (p. 49). However, this trend problematically leads to a form of "identity tourism" by privileged people who confuse their immersive experience with the reality of people from these groups (p. 54). In addition, these simulations are experiential practices that are limited in their political effectiveness, as they focus on individual experiences rather than on systemic changes that fight inequalities.

It is important to consider, moreover, how the viewpoints of underprivileged groups are portrayed. The types of simulations fostered by social VR projects can range, for example, from the experiences of refugees, people with disabilities, individuals in prison or solitary confinement, the homeless, factory-farmed animals, to the perspective of people of other different races or genders. Thus, the issue of ethical representation is crucial (Brey, 1999). It is worth asking who represents whom, especially when it comes to creating experiences that want to bring the standpoint of marginalized collectives or stigmatized identities to the fore. A key recommendation is then to involve the underprivileged groups that are represented in the development of these virtual simulations (Rueda & Lara, 2020). Such participation would help to mitigate concerns that these simulations produce misrepresentations and inculcate false beliefs, especially when they are intended to convey the first-person experiences of members of intersectional out-groups (see Ramirez et al., 2021).

Ultimately, this objection reminds us of the importance of the power relations that underlie the different parties involved in the socio-moral projects of VR. This factor is relevant, certainly, when the production and consumption of seemingly laudable projects that seek to combat various social injustices are carried out by privileged collectives.

The Real Perils of Bad Virtual Actions

For some people, the possibility of being able to abandon everyday morals may be an attractive feature of VR. Assaulting, stealing, killing, or raping are actions available in some virtual environments. For others, the above virtual experiences are examples of unacceptably risky content. In any case, the possibility of virtually performing courses of action that would generally be condemnable in non-virtual environments raises at least two interesting ethical questions.

On the one hand, the ethical status of virtual actions is a question to be asked (Brey, 1999). Is it wrong to treat virtual avatars badly? Similarly, which virtual behavior with morally reprehensible content—such as murder, rape, or pedophilia—should not be allowed? This has led to interesting ethical discussions, for example, on how to justify violent video games in which killing is a common practice that generates little scandal, as opposed to other behaviors that cause greater rejection, such as virtual pedophilia—the so-called "gamer's dilemma" (Luck, 2009).[6]

On the other hand, we may wonder whether bad virtual actions may lead to bad real-world actions or even to the development of an evil moral character. This is the issue that interests us the most here. Consider the case of violent VR games. Does virtual violence increase violent behavior offline? This is a typical concern in computer and gaming ethics. If violent VR games desensitize individuals to certain types of aggressive behavior, violent conduct by these persons may increase in the non-virtual world. Of course, this is an empirical question on which there is no clear evidence (Spiegel, 2018, p. 1542). However, it is often argued that this possibility may be harmful in terms of moral development, relying on an analogy with a typical Kantian argument. According to Philip Brey (1999, p. 9), "if disrespectful treatment of animals causes disrespectful treatment of human beings, then disrespectful treatment of virtual characters, which may be even more similar to such treatment of real humans, will have the same consequence". Furthermore, this could be more problematic with more advanced VR, where we may encounter more realistic gaming characters with high fidelity in human-like appearances (Kade, 2016, p. 82; Slater et al., 2020).

As it can be seen, this objection depends on an empirical premise, for which we do not yet have solid evidence. If it were shown that objectionable virtual behaviors do not translate into an increase in objectionable offline behaviors, this objection would not hold. Yet, the uncertainty does not allow us to dismiss this concern entirely. The long-term development of undesirable moral traits is a relevant worry as regards VR (Ramirez & LaBarge, 2020). The absence of long-term longitudinal studies on the effects of VR exposure (Madary & Metzinger, 2016; Slater, 2021) is an added motivation for this type of research—which may resolve the issue of this objection's validity in the future.

Concerns About the Well-Being of VR Users

Another factor that would diminish enthusiasm of employing VR for moral improvement would arise if virtual experiences reduce the users' well-being. There are various ways in which VR could go against users' interests. Broadly speaking, virtual experiences may cause psychological and health problems for users. This risk is especially prominent in cases in which extended immersion may lead to neglecting the physical environment, fostering depersonalization—i.e., seeing one's own physical body as strange— losing interest in the non-virtual world, increasing long-term social isolation, and even leading to bodily and child neglect (Madary & Metzinger, 2016; Spiegel, 2018; Slater et al., 2020). Moreover, in addition to being potentially addictive, the use of VR is sometimes accompanied by other addictive behaviors—especially the consumption of psychotropic drugs during gaming (Lorenz, 2020).

All in all, although we need more evidence on the after-effects of virtual simulations and the lasting consequences of long-term exposure, there is the risk of creating bodily and psychological harm. This is even more problematic when VR exploits the vulnerabilities of consumers such as teenagers, people with addictive tendencies, or individuals with mental disorders.

Another important interest of VR users has to do with privacy and personal data generated by their interactions with this technology (Madary & Metzinger, 2016; O'Brolcháin et al., 2016; Slater et al., 2020). Many VR applications are produced by platforms that may have a commercial interest in collecting user data. Therefore, the correct protection of personal data is key in order not to unduly threaten the privacy of users, especially when carrying out more sensitive behaviors in the virtual world. This concern is even more important if the use of VR systems that incorporate eye-tracking devices or emotion-capture technologies becomes widespread (O'Brolcháin et al., 2016).

Furthermore, there are issues specifically related to the use of VR in morally salient scenarios. This problem arises, in particular, when VR places users in immersive situations that stand out as morally demanding. There is a very interesting example from psychological research on moral behavior in virtual environments. Kathryn Francis and colleagues (2016, 2017, 2018) simulated the footbridge version of the trolley dilemma (Foot, 1967), in which the only way to save the five innocent people trapped on the rails from a runaway trolley is by pushing a large person to die from a bridge (Thomson, 1985). Whereas previous research had already simulated the lever version of the trolley dilemma (Navarrete et al., 2012; Patil et al., 2014; Skulmowski et al., 2014)—and even trolley-like scenarios reproducing traffic dilemmas with autonomous cars (Sütfeld et al., 2017; Faulhaber et al., 2019)— Erick Ramirez and Scott LaBarge (2020) found the footbridge version experiments especially controversial. In their view, the vividness of the later sacrificial moral dilemma may have generated stress and even trauma in the participants. Although elsewhere I have tried to downplay these concerns (Rueda, 2021), those risks should not be completely underestimated in the future.

According to Ramirez, it would be unethical in research contexts to place participants in experiences that we would consider unacceptable in the non-virtual world, what he calls "the principle of equivalence" (Ramirez, 2019). That idea is interesting beyond research and scientific domains, though. Would it be problematic to place users in stressful situations but from which we could obtain a socio-moral benefit? Would it be ethically impermissible? To answer this last question, the magnitude of the risks and benefits should first be sized.

Causing lasting psychological damage would be arguably unacceptable if the induced moral changes are minor, short-lived, or highly contextual. But if, on the contrary, the psychological costs are small and the benefits are high in terms of attitudinal and behavioral changes, we could be dealing with permissible simulations. Consider, for instance, the model proposed by Mel Slater and Domna Banakou (2021) about the Golden Rule Embodiment Paradigm for promoting helping behavior and diminishing antisocial behavior. This paradigm for using virtual embodiment to promote prosocial behavior is described, according to those authors, as follows:

> First, participants must be complicit in an action that causes harm to another person. Second, later, they must reexperience that episode from the embodied viewpoint of the victim, being able to observe their own previous actions (or acquiescence) during the course of the harmful event from that viewpoint.
>
> (Slater & Banakou, 2021)

These kinds of virtual experiences help us to put ourselves in the position of others, to avoid treating them in a way that we would not want to be treated, or to treat them like we would like to be treated—as the famous golden rule in moral philosophy recommends. To be sure, experiencing the victim's point of view, and perhaps also being the perpetrator, can be unpleasant for users. But if these interventions are effective in improving prosocial behavior, that distress may be a minor cost justifiable by the resulting major benefits. As mentioned in the second section, VR (as much as it may create transient discomfort) can help in experiential moral learning, by training us in the development of moral character. To borrow a popular saying, in morality, as in life, a smooth sea never made a good sailor.

Concluding Thoughts

Doing good with VR is a respectable aspiration, but not without its difficulties. Virtual environments and virtual embodiment generate opportunities for simulating experiences that help improve our moral tendencies. In this chapter, I have shown three areas where there may be particular—albeit sometimes contested—potential. VR can help foster empathic skills, reduce implicit biases, and enhance pro-environmental behaviors.

However, enthusiasm for using VR to improve human morality should be restrained. Many of the socio-moral projects may be comforting to privileged individuals and collectives, but the extent to which they improve the lives of the most disadvantaged must be studied. It is also worth considering the negative impacts that morally problematic virtual actions and risky content may have on the development of moral character. Finally, the well-being of users must be protected, trying to minimize the adverse effects that virtual experiences may have on their mental and bodily health, in addition to safeguarding the privacy of sensitive personal data.

I hope that the analysis offered in this chapter will encourage future research to fill the alarming gap concerning VR's long-term effects. As shown, empirical evidence is crucial for weighing the (beneficial or detrimental) impacts of VR from an ethical perspective. Consequently, not only should moral philosophers and applied ethicists be heartened to enter this debate, but scientists should also be encouraged to generate the strongest possible evidence. Whether VR becomes an ally for social good is an exciting topic for which there are still many unresolved questions.

Notes

1 The rhetoric of "good" has been at the forefront of various VR projects. For instance, the company Oculus, which belongs to Facebook, now rebranded as Meta, runs the program "VR for Good". This specific program promotes and

funds immersive journalism aiming to create progressive social impact. (See https://about.meta.com/community/vr-for-good/; last access 12th May 2023). Various immersive films for social awareness have also been cataloged under the name of the "'co-presence for good' movement" (Nakamura, 2020, p. 56).

2 Beyond computer-generated environments, 360° recorded videos that are projected on head-mounted displays can also be considered VR experiences, although these are generally less immersive than digitally designed environments (Slater & Sanchez-Vives, 2016, pp. 35–37). Later I will show that some of the hopes and criticisms of using VR to improve socio-moral behavior have also been focused on these immersive videos, especially regarding empathy enhancement. Therefore, in this chapter, I will manage a broad conception of VR that can accommodate different developments and applications of virtually immersive technologies.

3 See iAnimal360 of Animal Equality [Available at: https://ianimal360.com/; last access 31st January 2023] or the projects of People for the Ethical Treatment of Animals such as *I, Chicken, I, Orca*, and *Eye to Eye*.

4 The term "homuncular flexibility" refers to the high malleability of the body schema in VR, which facilitates even the virtual embodiment in avatars that do not have human appearance (Won et al., 2015a, 2015b).

5 It should be noted, however, that being embodied in an avatar representing a particular group is not equivalent to having the same actual experience as a member of that group (Lara & Rueda, 2021). Moreover, in some cases, the risk of misrepresentation is high. In disability simulations, for example, the participant may feel the temporary deprivation of a capacity more acutely than the disabled person who is accustomed to his or her condition and who does not have the same negative perception. Although some disability simulations (not through VR) may increase empathy towards that group, they may also provide misleading information and increase discrimination in subtle ways (Silverman, 2015). I am grateful to Andrew T. Kissel for bringing this point to my attention.

6 For ethical arguments that would allow virtual pedophilia through computer-generated child pornography (and also child-like sex robots), see Moen and Sterri (2018).

References

Agar, N. (2010). Enhancing genetic virtue? *Politics and the Life Sciences*, *29*(1), 73–75.

Aguiar, F., Gaitán, A., & Viciana, H. (2020). *Una Introducción a la* Ética Experimental [An Introduction to Experimental Ethics]. Madrid: Cátedra.

Ahn, S. J., Bailenson, J. N., & Park, D. (2014). Short- and long-term effects of embodied experiences in immersive virtual environments on environmental locus of control and behavior. *Computers in Human Behavior*, *39*, 235–245. https://doi.org/10.1016/j.chb.2014.07.025

Ahn, S. J. G., Bostick, J., Ogle, E., Nowak, K. L., McGillicuddy, K. T., & Bailenson, J. N. (2016). Experiencing nature: Embodying animals in immersive virtual environments increases inclusion of nature in self and involvement with nature. *Journal of Computer-Mediated Communication*, *21*(6), 399–419. https://doi.org/10.1111/jcc4.12173.

Ahn, S. J., Le, A. M. T., & Bailenson, J. (2013). The effect of embodied experiences on self-other merging, attitude, and helping behavior. *Media Psychology, 16*(1), 7–38.

Archer, D., & Finger, K. (2018). *Walking in Another's Virtual Shoes: Do 360-Degree Video News Stories Generate Empathy in Viewers?* New York: Tow Center for Digital Journalism, Columbia University. https://doi.org/10.7916/D8669W5C

Bailenson, J. (2018). *Experience on Demand: What Virtual Reality is, How it Works, and What it Can Do.* New York: WW Norton & Company.

Bailey, J. O., Bailenson, J. N., Flora, J., Armel, K. C., Voelker, D., & Reeves, B. (2015). The impact of vivid messages on reducing energy consumption related to hot water use. *Environment and Behavior, 47*(5), 570–592.

Banakou, D., Hanumanthu, P. D., & Slater, M. (2016). Virtual embodiment of white people in a black virtual body leads to a sustained reduction in their implicit racial bias. *Frontiers in Human Neuroscience, 601.*

Bang, E., & Yildirim, C. (2018). Virtually empathetic?: Examining the effects of virtual reality storytelling on empathy. In *Virtual, Augmented and Mixed Reality: Interaction, Navigation, Visualization, Embodiment, and Simulation: 10th International Conference, VAMR 2018, Held as Part of HCI International 2018, Las Vegas, NV, USA, July 15-20, 2018, Proceedings, Part I 10* (pp. 290–298). Springer International Publishing.

Batson, C. D. (2009). These things called empathy: Eight related but distinct phenomena. In J. Decety & W. Ickes (eds.), *The Social Neuroscience of Empathy* (pp. 3–16). Cambridge, MA: MIT Press.

Batson, C. D., Early, S., & Salvarani, G. (1997). Perspective taking: Imagining how another feels versus imaging how you would feel. *Personality and Social Psychology Bulletin, 23*(7), 751–758.

Batson, C. D., Lishner, D. A., & Stocks, E. L. (2015). The empathy-altruism hypothesis. In D. A. Schroeder & W. G. Graziano (eds.), *The Oxford Handbook of Prosocial Behavior* (pp. 259–281). Oxford: Oxford University Press.

Bedder, R. L., Bush, D., Banakou, D., Peck, T., Slater, M., & Burgess, N. (2019). A mechanistic account of bodily resonance and implicit bias. *Cognition, 184,* 1–10.

Bertrand, P., Guegan, J., Robieux, L., McCall, C. A., & Zenasni, F. (2018). Learning empathy through virtual reality: multiple strategies for training empathy-related abilities using body ownership illusions in embodied virtual reality. *Frontiers in Robotics and AI, 26.*

Bloom, P. (2016). *Against Empathy.* London: Bodley Head.

Bourzac, K. (2016). Neurostimulation: Bright sparks. *Nature, 531*(7592), S6–S8.

Brey, P. (1999). The ethics of representation and action in virtual reality. *Ethics and Information Technology, 1*(1), 5–14. https://doi.org/10.1023/A:1010069907461

Brownstein, M. (2019). Implicit bias. In E. N. Zalta (ed.), *The Stanford Encyclopedia of Philosophy* (Fall 2019 Edition). Stanford, CA: Stanford University. Available at: https://plato.stanford.edu/cgi-bin/encyclopedia/archinfo. cgi?entry=implicit-bias [last access 2nd February 2023].

Burkart, J. M., Brügger, R. K., & Van Schaik, C. P. (2018). Evolutionary origins of morality: Insights from non-human primates. *Frontiers in Sociology, 3,* 17.

Chowdhury, T. I., Ferdous, S. M. S., & Quarles, J. (2019). VR disability simulation reduces implicit bias towards persons with disabilities. *IEEE Transactions on Visualization and Computer Graphics*, 27(6), 3079–3090.

Cleveland, M., Kalamas, M., & Laroche, M. (2005). Shades of green: Linking environmental locus of control and pro-environmental behaviors. *Journal of Consumer Marketing*, 22(4), 198–212.

Coplan, A. (2011). Understanding empathy: Its features and effects. In P. Goldie & A. Coplan (eds.), *Empathy: Philosophical and Psychological Perspectives* (pp. 3–18). New York: Oxford University Press.

Cotton, M. (2021). The ethical dimensions of virtual reality. In *Virtual Reality, Empathy and Ethics* (Chapter 2, pp. 23–41). Cham: Springer. https://doi.org/10.1007/978-3-030-72907-3

Crockett, M. J. (2014). Moral bioenhancement: A neuroscientific perspective. *Journal of Medical Ethics*, 40(6), 370–371.

Crockett, M. J. (2016). Morphing morals: Neurochemical modulation of moral judgment and behavior. In M. S. Liao (ed.), *Moral Brains: The Neuroscience of Morality* (pp. 237–245). New York: OUP.

Damasio, A. R. (1994). *Descartes' Error: Emotion, Reason, and the Human Brain*. New York: G.P. Putnam.

Daniels, N. (2000). Normal functioning and the treatment-enhancement distinction. *Cambridge Quarterly of Healthcare Ethics*, 9(3), 309–322. https://doi.org/10.1017/s0963180100903037

Davis, M. H. (1980). A multidimensional approach to individual differences in empathy. *Catal. Select. Doc. Psychol. 10*, 85.

Davis, M. H. (1983). Measuring individual differences in empathy: Evidence for a multidimensional approach. *Journal of Personality and Social Psychology*, 44(1), 113.

DeGrazia, D. (2014). Moral enhancement, freedom, and what we (should) value in moral behaviour. *Journal of Medical Ethics*, 40(6), 361–368.

Di Nuzzo, C., Ferrucci, R., Gianoli, E., Reitano, M., Tedino, D., Ruggiero, F., & Priori, A. (2018). How brain stimulation techniques can affect moral and social behaviour. *Journal of Cognitive Enhancement*, 2, 335–347.

Douglas, T. (2008). Moral enhancement. *Journal of Applied Philosophy*, 25(3), 228–245.

Douglas, T. (2013). Moral enhancement via direct emotion modulation: A reply to John Harris. *Bioethics*, 27(3), 160–168.

Douglas, T. (2019). Enhancement and desert. *Politics, Philosophy & Economics*, 18(1), 3–22.

Douglas, T., & Devolder, K. (2013). Procreative altruism: Beyond individualism in reproductive selection. *Journal of Medicine and Philosophy*, 38(4), 400–419.

Dworazik, N., & Rusch, H. (2014). A brief history of experimental ethics. In C. Luetge, H. Rusch & M. Uhl (eds.), *Experimental Ethics. Toward an Empirical Moral Philosophy* (pp. 38–56). Cham: Springer.

Earp, B. D., Douglas, T., & Savulescu, J. (2018). Moral neuroenhancement. In L. S. M. Johnson & K. S. Rommelfanger (eds.), *The Routledge Handbook of Neuroethics* (pp. 166–184). New York: Routledge.

Eerp, B.D., Sandberg, A., Kahane, G., & Savulescu, J. (2014). When is diminishment a form of enhancement? Rethinking the enhancement debate in biomedical ethics. *Frontiers in Systems Neuroscience, 8,* Article No. 12.

Farmer, H., & Maister, L. (2017). Putting ourselves in another's skin: Using the plasticity of self-perception to enhance empathy and decrease prejudice. *Social Justice Research, 30*(4), 323–354. https://doi.org/10.1007/s11211-017-0294-1

Faulhaber, A. K., Dittmer, A., Blind, F., Wächter, M. A., Timm, S., Sütfeld, L. R., ... & König, P. (2019). Human decisions in moral dilemmas are largely described by utilitarianism: Virtual car driving study provides guidelines for autonomous driving vehicles. *Science and Engineering Ethics, 25*(2), 399–418.

Faust, H. S. (2008). Should we select for genetic moral enhancement? A thought experiment using the MoralKinder (MK+) haplotype. *Theoretical Medicine and Bioethics, 29,* 397–416.

Felnhofer, A., Kothgassner, O. D., Schmidt, M., Heinzle, A. K., Beutl, L., Hlavacs, H., & Kryspin-Exner, I. (2015). Is virtual reality emotionally arousing? Investigating five emotion inducing virtual park scenarios. *International Journal of Human-Computer Studies, 82,* 48–56.

Fisher, J. A. (2017). Empathic actualities: Toward a taxonomy of empathy in virtual reality. In N. Nunes, I. Oakley & V. Nisi (eds.), *Interactive Storytelling. ICIDS 2017. Lecture Notes in Computer Science* (vol. 10690, pp. 233–244). Cham: Springer.

Francis, K. B., Gummerum, M., Ganis, G., Howard, I. S., & Terbeck, S. (2018). Virtual morality in the helping professions: Simulated action and resilience. *British Journal of Psychology, 109*(3), 442–465.

Francis, K. B., Howard, C., Howard, I. S., Gummerum, M., Ganis, G., Anderson, G., & Terbeck, S. (2016). Virtual morality: Transitioning from moral judgment to moral action? *PloS One, 11*(10), e0164374.

Francis, K. B., Terbeck, S., Briazu, R. A., Haines, A., Gummerum, M., Ganis, G., & Howard, I. S. (2017). Simulating moral actions: An investigation of personal force in virtual moral dilemmas. *Scientific Reports, 7*(1), 1–11.

Foot, P. (1967). The problem of abortion and the doctrine of double effect. *Oxford Review, 5,* 5–15.

Gazzaniga, M. S. (2005). *The Ethical Brain.* New York: Dana Press.

Giubilini, A., & Savulescu, J. (2018). The artificial moral advisor. The "ideal observer" meets artificial intelligence. *Philosophy & Technology, 31*(2), 169–188.

Greene, J. D., Sommerville, R. B., Nystrom, L. E., Darley, J. M., & Cohen, J. D. (2001). An fMRI investigation of emotional engagement in moral judgment. *Science, 293*(5537), 2105–2108.

Groom, V., Bailenson, J. N., & Nass, C. (2009). The influence of racial embodiment on racial bias in immersive virtual environments. *Social Influence, 4*(3), 231–248.

Gyngell, C., & Selgelid, M. J. (2016). Human enhancement: Conceptual clarity and moral significance. In Steve Clarke, Julian Savulescu, C. A. J. Coady, Alberto Giubilini & Sagar Sanyal (eds.), *The Ethics of Human Enhancement: Understanding the Debate* (pp. 111–126). Oxford: OUP.

Hamilton-Giachritsis, C., Banakou, D., Garcia Quiroga, M., Giachritsis, C., & Slater, M. (2018). Reducing risk and improving maternal perspective-taking and empathy using virtual embodiment. *Scientific Reports, 8*(1), 1–10.

Harlow, J. M. (1848). Passage of an iron rod through the head. *The Boston Medical and Surgical Journal, 39*(20), 389–393.

Hasler, B. S., Spanlang, B., & Slater, M. (2017). Virtual race transformation reverses racial in-group bias. *PloS One, 12*(4), e0174965.

Heater, C. (1992). Being there: The subjective experience of presence. *Presence, 1*(2), 262–271.

Herrera, F., Bailenson, J., Weisz, E., Ogle, E., & Zaki, J. (2018). Virtual reality homeless. *PLoS One, 13*(10), e0204494.

Illes, J. (2003). Neuroethics in a new era of neuroimaging. *AJNR: American Journal of Neuroradiology, 24*(9), 1739.

Kade, D. (2016). Ethics of virtual reality applications in computer game production. *Philosophies, 1*(1), 73–86. https://doi.org/10.3390/philosophies1010073

Klincewicz, M. (2016). Artificial intelligence as a means to moral enhancement. *Studies in Logic, Grammar and Rhetoric, 48*(1), 171–187. https://doi.org/10.1515/slgr2016-0061.

Kraut, R. (2020). Altruism. In E. N. Zalta (ed.), *The Stanford Encyclopedia of Philosophy* 2016 (Fall 2020 Edition). Stanford, CA: Stanford University. Available at: https://plato.stanford.edu/entries/altruism/ [last access 31st January 2023].

Kulawska, A., & Hauskeller, M. (2018). Moral enhancement and climate change: Might it work? *Royal Institute of Philosophy Supplements, 83*, 371–388.

Lara, F. (2017). Oxytocin, empathy and human enhancement. *THEORIA. Revista de Teoría, Historia y Fundamentos de la Ciencia, 32*(3), 367–384.

Lara, F., & Deckers, J. (2020). Artificial intelligence as a socratic assistant for moral enhancement. *Neuroethics, 13*(3), 275–287.

Lara, F., & Rueda, J. (2021). Virtual reality not for "being someone" but for "being in someone else's shoes": Avoiding misconceptions in empathy enhancement. *Frontiers in Psychology, 12*, 3674. https://doi.org/10.3389/fpsyg.2021.741516

Levy, N., Douglas, T., Kahane, G., Terbeck, S., Cowen, P. J., Hewstone, M., & Savulescu, J. (2014). Are you morally modified? The moral effects of widely used pharmaceuticals. *Philosophy, Psychiatry, & Psychology, 21*(2), 111.

Ligthart, S., Meynen, G., Biller-Andorno, N., Kooijmans, T., & Kellmeyer, P. (2022). Is virtually everything possible? The relevance of ethics and human rights for introducing extended reality in forensic psychiatry. *AJOB Neuroscience, 13*(3), 144–157.

Lorenz, M. (2020). Commentary: The ethics of realism in virtual and augmented reality. *Frontiers in Virtual Reality, 1*(1), 113.

Luck, M. (2009). The gamer's dilemma: An analysis of the arguments for the moral distinction between virtual murder and virtual paedophilia. *Ethics and Information Technology, 11*(1), 31–36.

Madary, M., & Metzinger, T. K. (2016). Real virtuality: A code of ethical conduct. Recommendations for good scientific practice and the consumers of VR-technology. *Frontiers Robotics AI, 3*(February), 1–23. https://doi.org/10.3389/frobt.2016.00003

Maister, L., Slater, M., Sanchez-Vives, M. V., & Tsakiris, M. (2015). Changing bodies changes minds: owning another body affects social cognition. *Trends in Cognitive Sciences, 19*(1), 6–12.

Masto, M. (2015). Empathy and its role in morality. *The Southern Journal of Philosophy, 53*(1), 74–96.

May, J. (2011). Psychological Egoism. *Internet Encyclopedia of Philosophy*. Available at: https://iep.utm.edu/psychological-egoism/ [last access 31st January 2023].

Mendez, M. F. (2009). The neurobiology of moral behavior: Review and neuropsychiatric implications. *CNS Spectrums, 14*(11), 608–620.

Millar, H. (2016). Can virtual reality emerge as a tool for conservation? *Yale Environment 360*. 26th June 2016. Available at: https://e360.yale.edu/features/can_virtual_reality_emerge_as_a_tool_for_conservation [last access 28th January 2023].

Milk, C. (2015). How virtual reality can create the ultimate empathy machine. *TED Talk*. Available online at: https://www.ted.com/talks/chris_milk_how_virtual_reality_can_create_the_ultimate_empathy_machine [last access 31st January 2023].

Moen, O. M., & Sterri, A. B. (2018). Pedophilia and computer-generated child pornography. In B. David (ed.), *The Palgrave Handbook of Philosophy and Public Policy* (pp. 369–381). Cham: Palgrave Macmillan.

Nakamura, L. (2020). Feeling good about feeling bad: Virtuous virtual reality and the automation of racial empathy. *Journal of Visual Culture, 19*(1), 47–64. https://doi.org/10.1177/1470412920906259

Navarrete C. D., McDonald, M. M., Mott, M. L., & Asher, B. (2012). Virtual morality: Emotion and action in a simulated three dimensional trolley problem. *Emotion, 12*(2), 364–370.

Nozick, R. (1974). *Anarchy, State, and Utopia*. New York: Basic Books.

O'Brolcháin, F., Jacquemard, T., Monaghan, D., O'Connor, N., Novitzky, P., & Gordijn, B. (2016). The convergence of virtual reality and social networks: Threats to privacy and autonomy. *Science and Engineering Ethics, 22*(1), 1–29. https://doi.org/10.1007/s11948-014-9621-1.

Oh, S. Y., Bailenson, J., Weisz, E., & Zaki, J. (2016). Virtually old: Embodied perspective taking and the reduction of ageism under threat. *Computers in Human Behavior, 60*, 398–410.

Oh, S. Y., & Bailenson, J. (2017). Virtual and augmented reality. In P. Rössler, C. A. Hoffner & L. Zoonen (eds.), *The International Encyclopedia of Media Effects* (pp. 1–16). Hoboken, NJ: John Wiley & Sons.

O'Neill, E., Klincewicz, M., & Kemmer, M. (2022). Ethical issues with artificial ethics assistants. In C. Véliz (ed.), *The Oxford Handbook of Digital Ethics* (pp. C17.S1–C17.N26). Oxford: Oxford University Press. https://doi.org/10.1093/oxfordhb/9780198857815.013.17

Oxley, J. C. (2011). *The Moral Dimensions of Empathy. Limits and Application in Ethical Theory and Practice*. New York: Palgrave Macmillan.

Patil, I., Cogoni, C., Zangrando, N., Chittaro, L., & Silani, G. (2014). Affective basis of judgment-behavior discrepancy in virtual experiences of moral dilemmas. *Social Neuroscience, 9*(1), 94–107.Plechatá, A., Hielkema, M., Merkl,

L. M., Makransky, G., & Frøst, M. B. (2022). Fast Forward: Influencing the future in virtual reality increases real-life pro-environmental behaviour. *Preprint*. Available at https://assets.researchsquare.com/files/rs-2338078/v2/67fd5603-97eb-4a57-ba34-b5d1d4eb3969.pdf?c=1671652836 [last access 28th January 2023].

Plechatá, A., Morton, T., Perez-Cueto, F. J., & Makransky, G. (2022). A randomized trial testing the effectiveness of virtual reality as a tool for pro-environmental dietary change. *Scientific Reports*, *12*(1), 14315.

Peck, T. C., Seinfeld, S., Aglioti, S. M., & Slater, M. (2013). Putting yourself in the skin of a black avatar reduces implicit racial bias. *Consciousness and Cognition*, *22*(3), 779–787.

Persson, I., & Savulescu, J. (2012). *Unfit for the Future: The Need for Moral Enhancement*. Oxford: Oxford University Press.

Persson, I., & Savulescu, J. (2018). The moral importance of reflective empathy. *Neuroethics*, *11*(2), 183–193.

Prehn, K., & Heekeren, H. (2014). Moral brains—possibilities and limits of the neuroscience of ethics. In M. Christen, C. van Schaik, J. Fischer, M. Huppenbauer & C. Tanner (eds.), *Empirically Informed Ethics: Morality between Facts and Norms* (pp. 137–157). Cham: Springer.

Prinz, J. (2011a). Against empathy. *The Southern Journal of Philosophy*, *49*, 214–233.

Prinz, J. (2011b). Is empathy necessary for morality? In P. Goldie & A. Coplan (eds.), *Empathy: Philosophical and Psychological Perspectives* (pp. 211–229). New York: Oxford University Press.

Racine, E., Bar-Ilan, O., & Illes, J. (2005). fMRI in the public eye. *Nature Reviews Neuroscience*, *6*(2), 159–164.

Ramirez, E. (2017). Empathy and the limits of thought experiments. *Metaphilosophy*, *48*(4), 504–526.

Ramirez, E. J. (2019). Ecological and ethical issues in virtual reality research: A call for increased scrutiny. *Philosophical Psychology*, *32*(2), 211–233. https://doi.org/10.1080/09515089.2018.1532073

Ramirez, E. J., & LaBarge, S. (2018). Real moral problems in the use of virtual reality. *Ethics and Information Technology*, *20*, 249–263.

Ramirez, E. J., Elliott, M., & Milam, P. E. (2021). What it's like to be a _: why it's (often) unethical to use VR as an empathy nudging tool. *Ethics and Information Technology*, *23*(3), 527–542.

Ramirez, E. J., & LaBarge, S. (2020). Ethical issues with simulating the bridge problem in VR. *Science and Engineering Ethics*, *26*(6), 3313–3331.

Raus, K., Focquaert, F., Schermer, M., Specker, J., & Sterckx, S. (2014). On defining moral enhancement: A clarificatory taxonomy. *Neuroethics*, *7*(3), 263–273. https://doi.org/10.1007/s12152-014-9205-4

Read, H. (2019). A typology of empathy and its many moral forms. *Philosophy Compass*, *14*(10), e12623.

Rosenberg, R. S., Baughman, S. L., & Bailenson, J. N. (2013). Virtual superheroes: Using superpowers in virtual reality to encourage prosocial behavior. *PLoS ONE*, *8*(1), 1–9. https://doi.org/10.1371/journal.pone.0055003

Rueda, J. (2020). Climate change, moral bioenhancement and the ultimate mostropic. *Ramon Llull Journal of Applied Ethics*, *11*, 277–303.

Rueda, J. (2021). Socrates in the fMRI scanner: The neurofoundations of morality and the challenge to ethics. *Cambridge Quarterly of Healthcare Ethics*, 30(4), 606–612. https://doi.org/10.1017/S0963180121000074

Rueda, J. (2022). Hit by the virtual trolley: When is experimental ethics unethical? *Teorema*, 41(1), 7-27.

Rueda, J., & Dore-Horgan, E. (2022). A virtual prosthesis for morality? Experiential learning through XR technologies for autonomy enhancement of psychiatric offenders. *AJOB Neuroscience*, 13(3), 163–165. https://doi.org/10.1080/21507740.2022.2082584

Rueda, J., García-Barranquero, P., & Lara, F. (2021). Doctor, please make me freer: Capabilities enhancement as a goal of medicine. *Medicine, Health Care & Philosophy*, 24, 409–419. https://doi.org/10.1007/s11019-021-10016-5

Rueda, J., & Lara, F. (2020). Virtual reality and empathy enhancement: Ethical aspects. *Frontiers in Robotics and AI*, 7, 506984. https://doi.org/10.3389/frobt.2020.506984

Savulescu, J., & y Maslen, H. (2015). Moral enhancement and artificial intelligence: Moral AI? In J. Romportl, E. Zackova, y & J. Kelemen (eds.), *Beyond Artificial Intelligence. The Disappearing Human-Machine Divide* (pp. 79–95). New York, NY: Springer.

Schoeller, F., Bertrand, P., Gerry, L. J., Jain, A., Horowitz, A. H., & Zenasni, F. (2019). Combining virtual reality and biofeedback to foster empathic abilities in humans. *Frontiers in Psychology*, 9, 2741.

Schutte, N. S., & Stilinović, E. J. (2017). Facilitating empathy through virtual reality. *Motivation and Emotion*, 41, 708–712.

Schwartz, P. H. (2005). Defending the distinction between treatment and enhancement. *The American Journal of Bioethics*, 5(3), 17–19.

Seinfeld, S., Arroyo-Palacios, J., Iruretagoyena, G., Hortensius, R., Zapata, L. E., Borland, D., ... & Sanchez-Vives, M. V. (2018). Offenders become the victim in virtual reality: Impact of changing perspective in domestic violence. *Scientific Reports*, 8(1), 1–11. doi: 10.1038/s41598-01819987-7.

Silverman, A. M. (2015). The perils of playing blind: Problems with blindness simulation and a better way to teach about blindness. *Journal of Blindness Innovation and Research*, 5(2). https://nfb.org/images/nfb/publications/jbir/jbir15/jbir050201.html

Skulmowski, A., Bunge, A., Kaspar, K., & Pipa, G. (2014). Forced-choice decision-making in modified trolley dilemma situations: a virtual reality and eye tracking study. *Frontiers in Behavioral Neuroscience*, 8, 426.

Slater, M. (2009). Place illusion and plausibility can lead to realistic behaviour in immersive virtual environments. *Philosophical Transactions of the Royal Society B: Biological Sciences*, 364(1535), 3549–3557.Slater, M. (2021). Beyond speculation about the ethics of virtual reality: The Need for empirical results. *Frontiers in Virtual Reality*, 2(August), 1–4. https://doi.org/10.3389/frvir.2021.687609

Slater, M., Antley, A., Davison, A., Swapp, D., Guger, C., Barker, C., ... & Sanchez-Vives, M. V. (2006). A virtual reprise of the Stanley Milgram obedience experiments. *PloS One*, 1(1), e39.

Slater, M., & Banakou, D. (2021). The golden rule as a paradigm for fostering prosocial behavior with virtual reality. *Current Directions in Psychological Science*, 30(6), 503–509. https://doi.org/10.1177/09637214211046954

Slater, M., & Sanchez-Vives, M. V. (2016). Enhancing our lives with immersive virtual reality. *Frontiers in Robotics and AI, 3,* 74.

Slater, M., Gonzalez-Liencres, C., Haggard, P., Vinkers, C., Gregory-Clarke, R., Jelley, S., ... & Silver, J. (2020). The ethics of realism in virtual and augmented reality. *Frontiers in Virtual Reality, 1,* 1.

Sora-Domenjó, C. (2022). Disrupting the "empathy machine": The power and perils of virtual reality in addressing social issues. *Frontiers in Psychology, 13*(September), 1–13. https://doi.org/10.3389/fpsyg.2022.814565

Spiegel, J. S. (2018). The ethics of virtual reality technology: Social hazards and public policy recommendations. *Science and Engineering Ethics, 24*(5), 1537–1550. https://doi.org/10.1007/s11948-017-9979-y

Sundar, S. S., Kang, J., & Oprean, D. (2017). Being there in the midst of the story: How immersive journalism affects our perceptions and cognitions. *Cyberpsychology, Behavior, and Social Networking, 20*(11), 672–682.

Sütfeld, L. R., Gast, R., König, P., & Pipa, G. (2017). Using virtual reality to assess ethical decisions in road traffic scenarios: Applicability of value-of-life-based models and influences of time pressure. *Frontiers in Behavioral Neuroscience, 11,* 122.

Thomson, J. J. (1985). The trolley problem. *Yale Law Journal, 94*(6), 1395–1415.

Tomasello, M. (2018). Precís of a natural history of human morality. *Philosophical Psychology, 31*(5), 661–668.

Tsakiris, M. (2017). The multisensory basis of the self: From body to identity to others. *Quarterly Journal of Experimental Psychology, 70*(4), 597–609.

Van Loon, A., Bailenson, J., Zaki, J., Bostick, J., & Willer, R. (2018). Virtual reality perspective-taking increases cognitive empathy for specific others. *PloS One, 13*(8), e0202442.

van Schaik, C., Burkart, J.M., Jaeggi, A.V., & von Rohr, C.R. (2014). Morality as a biological adaptation—An evolutionary model based on the lifestyle of human foragers. In M. Christen, C. van Schaik, J. Fischer, M. Huppenbauer & C. Tanner (eds.), *Empirically Informed Ethics: Morality between Facts and Norms* (pp. 65–84). Cham: Springer.

Walker, M. (2009). Enhancing genetic virtue: A project for twenty-first century humanity? *Politics and the Life Sciences, 28*(2), 27–47.

Weinel, J., Cunningham, S., & Pickles, J. (2018). Deep subjectivity and empathy in virtual reality: a case study on the autism TMI virtual reality experience. In F. Michael & T. Veronika (eds.), *New Directions in Third Wave Human-Computer Interaction: Volume 1-Technologies*, 183–203, Cham: Switzerland.

Won, A. S., Bailenson, J., Lee, J., & Lanier, J. (2015a). Homuncular flexibility in virtual reality. *Journal of Computer-Mediated Communication, 20*(3), 241–259.

Won, A. S., Bailenson, J., & Lanier, J. (2015b). Homuncular flexibility: The human ability to inhabit nonhuman avatars. In R. A. Scott, S. M. Kosslyn & M. Buchmann (eds.), *Emerging Trends in the Social and Behavioral Science: An Interdisciplinary Searchable, and Linkable Resources* (pp. 1–16), Hoboken, NJ: John Wiley & Sons.

Yee, N., & Bailenson, J. (2007). The proteus effect: The effect of transformed self-representation on behavior. *Human Communication Research, 33*(3), 271–290.

Zaki, J. (2014). Empathy: A motivated account. *Psychological Bulletin, 140*(6), 1608.

11 Through a Glass Virtually
On the Future of Extended Realities

Erick José Ramirez and Andrew Kissel

In 1927, looking back on the success of the Model T, Henry Ford congratulated himself

> Nowadays everybody runs some kind of motor power but twenty years ago only the adventurous few could be induced to try an automobile. It had a harder time winning public confidence than the airplane has now. The Model T was a great educator in this respect. It had stamina & power. It was the car that ran before there were good roads to run on. It broke down the barriers of distance in rural sections, brought people of these sections closer together & placed education within the reach of everyone.
>
> (Ford 1927)

It might seem strange to compare Ford to Mark Zuckerberg or the mass production of cars to extended reality technologies but the comparison is apt. Aside from being among the richest people alive in their respective eras, they are also most famously associated with technologies they didn't invent but that went on to transform the world. The Model T helped create the car culture that led to the Interstate highway system, modern suburbia, fed the fossil fuel industry, and the assembly lines needed to create the T drove a massive restructuring of labor that uniquely affected the evolution of 20th-century capitalism. Meta, the company that owns Facebook, is betting over $100 billion (Gerstner October 24, 2022) that their extended reality (XR) headsets and virtual reality *Horizon Worlds* will change us all too.

History tends to repeat itself. Ford's Model T, as he admits, was not an immediate success. The infrastructure needed to support widespread use of cars didn't exist and the public was skeptical that the new technology was needed at all. Scholars of this era in American history note that there were "fierce debates about what motor vehicles ought to do and how they ought to fit into domestic life" (Wells 2007). The first Model T went on sale in 1908 and it took more than a decade before it came to dominate

DOI: 10.4324/9781003359494-15

the market. In these early years "the first American motor vehicles, owned almost exclusively by wealthy elites, represented a dizzying variety of designs that inspired spirited debate over such basic technological questions" (Wells 2007). Looking at the modern XR landscape today, looking back like this starts to feel a bit like looking into the future too. People aren't quite sure what to do with XR, how it ought to fit into domestic life, and what kind of infrastructure is needed to support it. The first commercial VR headset, the Oculus Rift, came to the market in 2016 followed shortly that same year by the HTC Vive.

Today it feels like everyone is writing off Meta's gamble to create the metaverse as a loss and the world of VR, more generally, looks to be struggling (Roth October 15, 2022). The Sony Corporation's investment into VR is, as of the time of writing, not paying off nearly as well as hoped (Bonthuys March 30, 2023). VRChat, the largest social XR application, averages less than 30,000 concurrent users (Hall July 28, 2022).[1] In the world of augmented reality, companies like Mojo Vision have decided not to bring their AR contact lenses to market despite spending years developing the technology (Greener January 17, 2023). In the end, *Horizon Worlds* and its immediate cousins might not take off but we shouldn't write off the idea of the metaverse.

The XR future envisioned by Meta, Sony, Mojo Vision, and VRChat Inc., one where billions of users live, work, and play in places that look and feel real is practically inevitable. Extended reality technologies are being researched and developed around the world by institutions with tremendous economic, political, and social capital. Microsoft continues to refine and develop its AR Hololens. The US Military is exploring combat uses for AR (Congressional Research Service 2022). Walmart, one of the largest retailers in the world, has invested heavily for years in VR training (Incao September 20, 2018). Chinese media giant Tencent has also recently begun to create its own mixed reality unit (Ye & Yang June 20, 2022), and technology giant Apple Inc. is set to debut its XR headset in 2023 (Pritchard April 21, 2023). There are too many global players invested too deeply in the potential of these technologies for them to depend on the success or failure of one social XR world or technology.

In this chapter, we look forward to a future where the infrastructure for the widespread adoption of XR technologies is in place and where the most pressing questions about XR are no longer the concerns we have today about affordability, access, safety, or privacy. The XR technologies being created today aren't going anywhere and we have good reason to think we're not ready legally, politically, morally, and even metaphysically, for the changes these technologies will unleash in the 21st century.

We'll center our analysis on two broad questions. The first, "How (can) XR change us?" explores what we propose is a radical shift in conceptions of the self that will challenge traditional ideas of separations between the physical and virtual. We then turn to questions of value, both ethical and aesthetic, and the new opportunities that XR proposes for individual exploration and expression. Our second question, "What are the social/ political implications of XR?," maps out new ethical questions about ourselves, our love and sex lives, how we organize, the role of XR overlays in physical spaces, and coming questions about XR bots and extended minds. To address these questions, we need a new ethics and we need it sooner than later. The technologies that have been (and are being) developed in service of the broad XR spectrum are becoming increasingly entrenched in commercial and industry applications. We may not know what the future of XR looks like, exactly, but there is still reason to think that underlying technologies, and the ways we interact with them, will continue to push our philosophical frameworks to the limits.

XR and Our Individual Futures

Individual Bodies

One of the proposed benefits of XR is that it allows users to present themselves through virtual bodies that are very unlike the physical bodies they inhabit every day. For example, "Traveling while Black," a VR documentary film about the risks of traveling in the United States in the 1960s as a black person, attempts to recreate the experience of searching for the safe spaces outlined in the Green Book traveler's guide (Schrader 2022). More than just a collection of interviews, "Traveling while Black" tries to give the user a taste of the kind of discrimination that many minorities faced at the time. The thinking seems to be that users will have greater empathy for someone else after they have experienced what it is like to be someone else in XR. Such "what it is like" claims (e.g., experience what it is like to be a…) abound in VR development circles.

Rueda (this collection; Chapter 10) argues that studies on the proposed benefits of this kind of "XR embodiment" are far from conclusive. More generally, we should be cautious in our application of the term "embodiment" to these kinds of experiences in the first place. Philosopher Maurice Merleau-Ponty holds that the organic body, and oneself as situated within a body within the world, is foundational to our subjective experience. Put another way, the conscious experience cannot be extrapolated away from our physical embodiment (Merleau-Ponty 1945 (2013)). In Merleau-Ponty's sense, then, XR experiences in which one presents themselves in a virtual body fail to count as "embodiment" in at least two ways. First, the sensory stimulation presented by many XR experiences fails to interact with the body as a whole. Sure, you may have visual and auditory stimuli that simulate non-virtual experiences. But olfactory, haptic, proprioceptive, and other feedbacks are missing. Furthermore, the way that your body can interact with the virtual environment is limited by current XR technology interfaces. Because the whole body is not engaged in a subject-world circle of causality, it fails to be an embodiment in Merleau-Ponty's sense. Second, because the user is still very much in their own physical body when they enter XR, they bring the totality of their previous lived experience with them as well. Spending 20 minutes experiencing the discrimination and oppression of traveling as a black person is not the same as discrimination being a constitutive component of one's ongoing lived experience. As such, there are serious limits to the claim that XR users get to experience "what it is like to be a…"

Such criticisms are not unique to XR. In an effort to draw attention to the needs of the visually impaired, some organizations hold "blindness simulations" where participants try to eat a meal or complete other mundane tasks while blindfolded (Silverman 2015). The problem with these kinds of simulations is that participants, having not been blind for very long, do not have any well-worked-out strategies for dealing with their (temporary) lack of sight. Their experience thus contrasts with a person whose daily life has evolved to function without vision. As a result, blindness simulations can have the effect of *reinforcing* negative stereotypes toward blind people. Putting on a blindfold is not the same as learning about the lived experience of what it is like to be blind.

Nevertheless, XR does seem to provide users with powerful ways of exploring their own bodily identity. For example, a trans woman who participated in a 2020 study on the perception of one's avatar in social VR said,

> Using a feminine avatar makes me confident not only in VR but also in real life. I feel like that would be actually more real than the real you in real life. Because in real life, you're stuck with what you were born with. But in VR, you can be what you truly feel like you are inside. This experience actually gave me confidence to start my [transgender] procedure in the real life.
>
> (Freeman et al. 2020, 5)

In this case, the user's experience in VR gave them such a strong sense of what it would be like to live in a more feminine body that they initiated steps to bring their physical bodies more in line with their virtual bodies.

These reflections suggest a tension on the near horizon for the exploration of bodily identity and XR embodiment. On the one hand, claims of XR embodiment should not be read too strongly (in the Merleau-Ponty sense), lest they reinforce discriminatory thinking and stereotyping. On the other hand, some users report such strong connections between themselves and their XR avatars that dismissing them as "mere" virtual overlays also seems misplaced.

The idea that we bring our whole (non-virtual) selves to be virtually embodied in XR experiences is increasingly strained. It's worth exploring the way that XR may allow us to create bodies that blend traditional distinctions. For this reason, we propose that further research should consider the possibility of what we call *body blending*. Genderqueer identities blur traditional distinctions between men and women and create entirely new gender categories (Dembroff 2020). Similarly, body blending would involve blurring virtual and non-virtual bodily identities.

While such blending may sound farfetched, we maintain that it may already be occurring. Consider the following case: Sam spends as much time in social VR as they do outside of it. In VR, Sam uses a feminine avatar, but outside of VR, Sam's physical body presents in stereotypically masculine ways. Similar to the trans woman quoted above, Sam finds that using a feminine avatar increases their confidence in VR and outside of it. Unlike the earlier trans woman, however, Sam feels no inclination to modify their non-virtual body to bring it more in line with their virtual body. As such, we could plausibly understand Sam as engaging in body blurring in at least two ways. First, we can understand Sam's concept of "body" as expanding to include both their virtual and non-virtual bodies. After all, Sam spends equal time interacting with environments through a virtual interface as without it. Second, Sam's lived experiences can be understood as the blending of their living in virtual and non-virtual spaces. Insofar as Sam brings the entirety of their lived experience in a physical male body to bear in virtual environments, it would be infelicitous to say that Sam learns "what it is like to be a woman" while in VR. However, Sam does seem to learn *something* about what it is like to present in a way that diverges from their presentations in other domains.

The picture sketched out by the case of Sam, and the possibility of body blending more generally, is in need of further investigation. Nevertheless, it does not seem outside the realm of plausibility that as people invest more time in virtual settings, perhaps even spending more time in virtual settings than outside of them, the physical restrictions placed on identity that underpin many traditional concepts, such as "body," "gender," and perhaps even "race," could become increasingly strained. Hence, body blending.

It's important to remember that embodiment and identity are greatly disputed even before the introduction of virtual embodiment and body blending. For example, despite the growing acceptance of transgender identities among philosophy communities, transracial identities remain fraught with controversy (Tuvel 2017; Cattien 2019). Indeed, there is broad division on how to analyze racial concepts in the first place (Mallon 2006, 2022). When the possibility of transgender and transracial identities intersects, the resultant "transgracial" identities are even more controversial (Gladden 2015). The possibility of body blending in virtual settings promises to make these disputes even murkier.

At this point, one may observe that the physical body is more fundamental than the virtual. Without a physical body, the thinking goes, virtual bodily identities are not even possible in the first place. As such, one's physical body always takes priority over any possible virtual embodiment. After all, it's always possible to log out of a virtual space in a way that one cannot escape the physical body they find themselves in.

In our view, although the physical body may be *causally* prior to the virtual body, it remains to be seen whether this priority extends to primacy in other domains. Previous chapters in this collection dispute how the metaphysics of XR are best understood, including one argument that metaphysical interpretations should be bypassed entirely (Silcox Chapter 2, Aarseth Chapter 1, Tavinor Chapter 3). There are further questions about how these interpretations would impact our understanding of our own bodies in virtual spaces. For example, why should we think that a physical body is more *real* than a virtual body? A virtual chessboard and a physical chessboard are both equally adequate for playing a game of chess. So plausibly, a virtual chessboard is a real chessboard. If correct, then the fact that one board is digital and the other is not does not make the games played in physical spaces more *real*. And it certainly doesn't make the games played in physical spaces more *meaningful*.

It remains to be seen whether arguments for the reality and meaningfulness of virtual chess will work as well for virtual bodies. It is worth noting once again how the possibility of virtual embodiment and body blending interacts with ongoing disputes about the nature of bodily identity more generally. For example, Aas (2021) argues that physical embodiment should be understood in a moralized way so as to include both organic parts as well as non-organic prostheses. Such a view could plausibly be modified and extended to include virtual bodies as well, though further research is obviously necessary.

It does seem like the fact that (most) people currently spend more time in non-virtual spaces than in virtual spaces impacts our answers to these questions. But as more time is spent in the virtual (which, if some corporations get their way, is in the near future for those who work from home), our attitudes toward what counts as one's body could begin to shift. During the COVID-19 pandemic, individuals spent increasing amounts of their social time interacting with each other through VR. Documentaries such as *Life 2.0* (2010) and *We Met in Virtual Reality* (Hunting 2022) depict thriving social communities that engage with each other in entirely virtual spaces. Friends recognize each other in these spaces based on the appearance of virtual avatars, much as one identifies friends in non-virtual settings through the appearance of their physical bodies.

Aesthetic modifications of one's virtual avatar may in turn come to be viewed in the same way as aesthetic modifications to one's physical body. In non-virtual spaces, heavy use of makeup and plastic surgery are used to augment how one appears to others without substantially undermining one's identity claims. As explored by Ramirez et al. (this collection Chapter 6), it is not clear why virtual filters and avatars should not be understood in similar ways. Perhaps in the future, "I love your new fox body," could be understood as a claim about *you* in much the way that the phrase, "I love your new green hair," is understood today.

Individual Actions & Value

To close out this section, we turn to questions of what one does with their virtual bodies. What values should govern virtual actions? What does an ethics of virtual actions look like? The issue here is not how we can use XR technologies ethically (for more on this question see Madary et al. 2016). Rather, the issue is to determine what constitutes ethical behavior for users within contained XR experiences. In the case of social XR experiences, previous concepts governing interactions between individuals may need to be ameliorated and expanded to account for new kinds of virtual interaction. For example, several researchers have explored the way in which sexual harassment manifests in social XR (Blackwell et al. 2019; Freeman et al. 2022; Ramirez et al. 2023). There is disagreement in these cases about what constitutes "harassment" in XR, even across victims. Further work must be done to determine what we owe each other in virtual contexts.

Questions regarding the ethics of virtual actions arise in non-social virtual experiences as well. These questions have been explored in a variety of manners through the ethics of video games (Tavinor 2009; Sicart 2011; Bartel 2022). A major driving force in this literature has been the attempt to resolve the so-called gamer's dilemma (Luck 2009). Briefly, the gamer's dilemma picks out the tension between the apparently widespread acceptance of violence in video games, on the one hand, and the widespread rejection of pedophilia in video games. Various proposals have been presented in the literature to resolve the dilemma, including identifying morally relevant differences between the two types of actions (Bartel 2012; Young 2016; Luck 2022) and reevaluating the apparent source of the dilemma (Ali 2015, 2022; Ramirez 2020).

More generally, the gamer's dilemma raises the question of what we ought to be doing in virtual worlds. On the one hand, since no actual people are harmed in non-social XR experiences, they seem to provide a safe place for experimentation with and exploration of ideas, identity, experiences, etc. that would not be possible in non-virtual settings. On the other hand, the events simulated in XR experiences often have real-world counterparts (Patridge 2011). As Heinrichs (2021) argues, virtual actions can be *representative* acts and should be morally evaluated on that basis. As such, a virtual murder may not be an *actual* murder, but insofar as it *represents* a murder, we may morally evaluate it in much the way we would morally evaluate a painter who depicts the murder of a particular individual in their work. Determining individual ethics of virtual actions will thus depend crucially on what we take virtual actions to be and how they are embedded in the broader culture, even in the case of non-social XR experiences.

As a closing remark on the individual ethics of virtual actions, we wish to consider what the good life looks like in XR. Earlier, we suggested that as people spend increasing amounts of time in XR, their bodily identities may begin to blend physical and virtual manifestations. Kissel (this collection, Chapter 7) suggests that one's narrative identity can be partially constituted by virtual narratives. If one is building one's identity, and living according to ethical principles, in XR, then what kind of XR life should one live?

One potential answer is that one should *not* live a life in XR. In 1974, Robert Nozick introduced the idea of the "experience machine," which would allow you to float in a tank connected to electrodes and experience the pleasure of a life well lived. At its core, Nozick's experience machine is nothing more than an IVR simulator.[2] Nozick famously argued that the vast majority of individuals would turn down the opportunity to plug into the experience machine for life (Nozick 1974, 44). For him, the value of a life lived inside the experience machine could not compare to the value of a life lived outside of it, since "we want to *do* certain things, and not just have the experience of doing them" (Nozick 1974, 43). One plausible extension of this thinking is that one cannot live the good life in XR, as the virtual actions one performs in XR lack a certain kind of value.

Whether one agrees with this extension obviously depends on how one characterizes "the good life." Nevertheless, we think it is a hard line to hold that virtual actions can *never* contribute to the good life. If reading a work of philosophy (plausibly) contributes to the good life, then using XR can as well, at least in principle, through the experiences and knowledge they make available to users. It is a further question, however, whether the value of XR experiences can only be understood with respect to the *non-virtual* life to which they contribute. Answering this question will depend on further exploration of the way that we value the virtual and its relationship to the non-virtual. Ali (2023) suggests assessing the value of the virtual along two main axes: a reproduction-simulation axis and a representation-simulacra axis. These axes allow Ali to argue that, to the extent that the virtual reproduces *essential* properties while representing *other* properties of their non-virtual counterparts, the virtual reproductions are valuable. Notice that this approach assumes that if the virtual alternatives are to be valuable, it will be for the same reasons that their non-virtual counterparts are valuable. The possibility that virtual entities, actions, events, etc. could be valuable for alternative remains is in need of further exploration.

If we reject the extension of Nozick's experience machine argument, what would a life well lived in a virtual world look like? Would it be analogous to the good life outside of XR? Would it be more or less permissive? Would it require social interactions? Answers to these questions will affect how we choose to spend our time in XR, but also what kinds of experiences we choose to build in the first place.

It is also worth considering questions of *aesthetic* value in virtual contexts. Perhaps unsurprisingly, a great deal of prior philosophical work on virtual worlds has focused on aesthetics (Robson 2018). One question of primary interest is whether interactive virtual worlds should be considered works of art. Although there is some popular support for the claim that virtual games are not art (e.g., Ebert 2010), the majority of theorists agree that there is no in principle barrier to creating art in the form of interactive virtual experiences (Smuts 2005; Tavinor 2009). These approaches have tended to look at historical examples of artworks, as well as theories of art, in order to establish that virtual worlds are similar to also count as artworks. For example, many virtual experiences use lens flare even though the "camera" in virtual experiences is entirely digital. In doing so, the experience adopts the language of earlier movies and films. Such arguments tend to place virtual worlds in the tradition of representational artforms such as photography, painting, and cinema (Tavinor 2021).

Other theorists, however, fear that virtual worlds are pigeonholed when placed alongside previous representational and narrative artforms for the purpose of aesthetic evaluation (Juul 2001; Aarseth Chapter 1 of this collection). Focusing particularly on virtual games, these theorists see game elements as constitutive of the aesthetic value of virtual worlds. For example, Thi Nguyen argues that games provide an opportunity to adopt temporary agencies and that the experience of agency is itself an aesthetic experience (Nguyen 2020). Others, like Robson (2018) maintain that the playing of video games can be understood as unique aesthetic performances, unlike previous forms of art.

Although these arguments focus on games, we see little reason to think that they could not comfortably extend to non-gaming virtual experiences.[3] As discussed above (and in previous chapters of this collection), entering an XR experience often involves temporarily adopting a virtual body, with temporary goals that may differ from the goals one has prior to entering the experience. Plausibly, then, XR experiences provide opportunities to explore temporary agencies. With such temporary agencies comes the possibility of new forms of aesthetic experience.

The possibility of agential and performative aesthetics in XR interacts with questions regarding values in XR explored above. There may be aesthetic value in XR experiences, not just from the perspective of the developer as the creator of a work, but from the perspective of the user exercising agency in virtual worlds. In turn, these new aesthetic horizons may inform our answers to ethical questions that arise in the development and use of XR experiences (Gigliotti 1995). For example, consider the "cozy games" subgenre of video games. Although there is some disagreement as the genre develops, a cozy game will be "…laid back, have minimal, if any, combat, an endearing art style, and will wrap its action around a wholesome story" (Bellingham 2022). Cozy games, such as *Animal Crossing*, use "strong aesthetic signals that tell players they are in a low stress environment full of abundance and safety" (Cook 2018). The abundance and safety experienced by users, in turn, allow users to adopt agencies in keeping with feminist philosophical and ethical themes (Waszkiewicz & Bakun 2020). In this way, the aesthetics of cozy games, including their gameplay, support broader ethical and value judgments. Once again, previous philosophical work has generally focused on the connection between aesthetics and ethics in the context of games. More research should be focused on how XR experiences more generally allow for new forms of aesthetic, and ethical, exploration.

XR and Our Social/Political Futures

XR body blending is a way of talking about our experiences of ourselves in relation to our XR bodies. Whether or not embodiment itself is the best way of thinking about these phenomena (as opposed to other frameworks like extended cognition or structural intersectional frames) is itself an interesting question. That question aside, XR body blending will, and in many ways is already beginning to, affect how individuals think of themselves, their autobiography, and the connection between the self and our physical and virtual bodies.

In the preceding section, we explored these not-too-distant philosophical questions about XR technologies and the body. While changes to the individual are important and will keep scholars busy for some time to come, XR technologies are likely to have impacts that extend beyond ourselves; they'll end up affecting our social and political landscape too. In this section we explore that landscape, fully aware of the fact that predictions of the future can often turn out to be not only wrong but also mere projections of present-day anxieties. We begin again with XR embodiment to explore its social and political implications before treading onto (increasingly) hypothetical territory with Deepfakes, informational overlays, harassment, and the rigid distinction between humans and human-like bots.

If XR bodies matter to us as individuals, and the evidence looks like they do (or at least...that they can and will), then control over those bodies is likely to constitute a significant social and political debate in the future. Ramirez et al. argued, in Chapter 6 of this volume, that the regulatory future for embodiment will be clouded by conflicting demands to (1) allow individuals to control how they appear to others, (2) allow individuals to control how others appear to them, (3) allow corporations to own and control embodiment within their metaverses, and (4) allow states to regulate XR body blending. While each may have some face plausibility, all options produce problems that will need to be resolved and the options are jointly inconsistent.

Regulatory questions notwithstanding, we might find our social norms changing in unusual ways as we enter a space where traditional assumptions about the physical body and the self are challenged or become less relevant. Historically, Western conceptions of romantic love have often been grounded on Platonic or Aristotelian traditions fixed, at least partially, by physical bodies (Plato 1989; Sherman 1993).[4] As such, our future theories of romantic love may need to stretch (without breaking) in order to accommodate XR embodiment. Can we love someone whose embodiment changes from meeting to meeting? Some readers, at this point, may argue that relationships in XR are bound to fail because of the limited opportunities for interaction that exist in social XR spaces because the distances between potential lovers are too great, or simply because sex is impossible.

Scholars have discussed the possibility of genuine online relationships for some time (Ben-Ze'ev 2004; Vallor 2012). XR technologies, and thus XR relationships, are comparatively less well understood. Authors in this volume (Madary Chapter 5, Parsons & Neisser Chapter 4) have called attention to important features of XR experience that bear upon XR relationships. It's becoming clearer, for example, that people can experience things in XR as if it were really happening to them. People in an XR space feel physically close to one another because their avatars are close to each other even if the physical distance between them is vast. Furthermore, they may feel as if they are able to physically interact both as a result of haptic feedback generated by the technology but also because of the expectation of the touch itself. This experience of physical touch without haptics has been referred to as "phantom touch" (Freeman 2022). Despite this, we may still feel like sex is important and yet remains impossible in XR environments.

> Romantic love isn't about sex (a common fallacy) but it depends on sex, thrives upon sex, utilizes sex as its medium, its language and often its primary content. Whatever else it may be, romantic love begins with the inspiration and exhilaration of sexual attraction...sexual attraction is not 'just physical' of course...but whatever else it may be, sex is bodily and sexual desire engages us as embodied creatures for whom 'looks' and the blessings of nature are at least as important as the egalitarian insistence that we are all, 'deep down,' essentially the same.
>
> (Solomon 1988)[5]

Theories that aim to turn love into a spiritual or psychological union may seem to fare better though even in these cases theorists often argue that love begins physically (even if we transcend physicality eventually). How might our conceptions of love and sex change in a world where embodiment is not merely physical, and dependent on the blessings of nature, but embodied and (in one analysis of how XR body blending should be regulated) entirely designed by the individual? Can we make sense of embodied sex that includes XR body blending? Guo Freeman and her colleagues have investigated how users can use XR avatars to explore and experiment with, central aspects of their identities. She argues that

> Avatars play a central role in the communicative dynamics in virtual worlds; they integrate several different social values such as gender roles and social norms. They also afford the experimentation of completely new identities (e.g., cross-gender play) or reaffirmation of existing identities (e.g., queerness gameplay).
>
> (Freeman 2020, 2)

In the same vein, we've inherited our conceptions of fidelity, monogamy, and affection from forebearers for whom physical embodiment was the only way to understand the self. We face a host of philosophical questions about whether to maintain, remove, or ameliorate these concepts in the light of XR body blending. We'll need to be ready to discuss how these new forms of relations and relationships should be recognized by the state and how relationship rights should be distributed. Things get even more complicated if, as we say at the end of this section, we end up spending more time with XR-embodied human-like bots.

While we hope it's clear that restructuring our ideas about love, relationships, fidelity, and monogamy will bring with it social and political change,

XR's influence on our future politics extends beyond these concerns. Take, for example, what researchers call "the proteus effect" (Yee & Bailenson 2007, Han & Bailenson in this collection Chapter 9). The proteus effect is a way of referring to the fact that people will tend to change how they behave so that it "conforms to their digital self-representation independent of how others perceive them" (Yee & Bailenson 2007). In the original studies supporting this effect, subjects were embodied in avatars that were taller than average or conventionally attractive; researchers found that their subjects began to take on behavior characteristics of individuals who really are taller than average or conventionally attractive. Some researchers have explored the bounds of the proteus effect when it comes to actual (or hypothetical) empathy enhancement by embodying their subjects in avatars coded with different racial or gender identities (Banakou, Hanumanthu, & Slater 2016; Ramirez, Elliott, Milam 2021).

If corporations have the power to shape embodiment, it's possible that more is at stake than our sense of self (important as that may be). The way that we experience the world can be shaped and altered, to suit the whims of the corporation controlling whatever XR environment we're inhabiting at any given time. We shouldn't presume that this sort of shaping, what philosophers of technology call "nudging" (Sunstein 2015) is always unethical. Our point in discussing it here, and now, is that we'll need an ethical and regulatory framework for these interventions and, ideally, we'll need to have this framework in place *before* we spend a lot of time in these spaces. Imagine, for example, a dystopian scenario in which our XR bodies (and those with whom we interact along with those who represent individuals in power) could be controlled so as to minimize (or maximize) in-group/out-group biases or to nudge individuals into focusing on interpersonal differences as a way to obfuscate the real sources of social/economic inequality or distress. While these are dangers already associated with misinformation on social and other online media spaces, Michael Madary, in his contribution to this volume, convincingly argues that XR technologies uniquely threaten our traditional ways of knowing in ways we are not yet ready (legally or conceptually) to fully appreciate.

Similarly, while data privacy has become a common concern among all technologically oriented discussions, the forms of data that can be captured by XR technologies raise concerns that our current regulatory and moral frameworks may not be ready to address (see also the contribution by Parsons & Neisser in this volume). While public consciousness has caught on to the importance of protecting sensitive data like passwords and health records, biometric data remains both difficult to legislate and to draw public concern about (Hargittai & Marwick 2016; Kugler 2019). In an era where consumers willingly trade information about their fingerprints in order to log onto their devices, share their locations with friends and with driving apps, and happily post photos on social media that allow corporations to store facial biometric data, it's even more important that we appreciate that the data security threat posed by XR is not just more of the same. Whatever consumers' attitudes are toward personally identifiable information (fingerprints, facial features, health records, etc.) such data is recognized and understood as deserving special protection. The data that can be collected by XR devices can go beyond this. Data that is not typically thought to be personally identifiable, like motion tracking recorded by XR handsets and hmds, can very quickly become personally identifiable. In some cases, collecting five minutes of a person's motion data is enough to positively identify them with 95% accuracy (Miller et al. 2020).

Inward-looking cameras on XR devices can track your eyes, record your facial expressions, and predict how much you're concentrating on whatever it is you're doing. Some VR headsets today can record EEG activity on your scalp to study your brain's responses to whatever you're doing and many others can also combine this data with biometric information like a user's pulse to draw inferences about your thoughts and feelings. Putting all these data together gives XR developers the power to make really intimate predictions about you. They can infer your desires, your thoughts, and your emotions by looking at how your body responds to XR content. Emteq labs, for example, touts that its product "incorporates biometric sensors, signal processing modules, gaze and optional eye-tracking, supported by a unique open API and cloud data analytics, tracking, management. Our solution provides insights into emotional state & response on demand" (emteq n.d.)This information is not only intimate but valuable. When you enter an XR simulation, companies today can record your avatar design choices, track every move you make, record everything you look at or interact with, and predict how those things make you feel.

Knowing this much about each of their customers gives XR developers a lot of control to feed their predictive algorithms and this data is a valuable commodity to advertisers who want to target people like you. They can even get near-instant feedback on how well their ads are working by recording your responses. Today, companies like Meta promise not to hold on to the images taken by these cameras but the *information* about these images can be kept. That information includes sensitive inferences they make about you and the way you feel. According to publicly available documents released by Meta, "if you have chosen to share additional data with Meta, we collect additional data about how you use your headset (including Natural Facial Expressions) to help Meta personalize your experiences and improve Meta Quest" (Natural Facial Expressions Privacy Notice 2022).

It might sound silly to care about whether Meta (or Sony or Tencent) can tell that you're smiling when you're virtually hanging out with a friend.

The now-defunct corporation, Cambridge Analytica, famously got into serious hot water during the run-up to the 2016 US elections and the UK Brexit resolution, for using detailed information about Facebook users to deliver highly targeted (and highly effective) advertisements to sway public opinion. Highly detailed information about our likes, dislikes, engagements, and shares was enough to allow Cambridge Analytica to construct highly manipulative, and morally questionable, advertisements. Some ethicists are sounding the alarm about how XR technologies have the potential to manipulate us in ways far deeper than Cambridge Analytica was able to (Spiegel 2018). Corporations already have access to vast psychographic information about us (our likes, dislikes, consumer preferences, socioeconomic level, where we live, what we post about, etc). When this information is added to real-time information about what we're looking at, who we're interacting with (in both physical and virtual spaces), and how we feel about our experiences as they happen to us, the opportunities for manipulation grow even larger. Such technologies can potentially "transform the structure of our life-world" giving corporations control over how we think and see the world to shape us into ideal consumers (Madary & Metzinger 2016).

Privacy advocates argue the immense scope of intimate data XR devices gather about you and the people who share a space with you requires a massive shift in how we think about data privacy (Dick 2021). We're already used to thinking about data privacy, the security of our browsing history, geolocation, and strong passwords. It might seem like XR privacy might be more of the same. But the data collected by XR is orders of magnitude greater and more intimate. They measure not only where you are but what your body is doing, what you're looking at, and who's in the room with you. Advocates already argue that we need new privacy protection policies to protect this expansive new data some have referred to as "biometric psychography" (Heller 2021).

Beyond questions about what data can be collected about *us* and what that data can be used for, XR poses unique risks because of the unique psychological responses we can have to XR content but also because of the relationships between XR and other technologies. For example, Deepfake technologies are designed to create convincing false auditory or visual representations. Because Deepfakes use actual audio, images, or video of the person being faked, a good Deepfake can easily fool someone into thinking that they're actually seeing or hearing someone they're not. In traditional online spaces, Deepfakes introduce a host of ethical issues relating to deception (de Ruiter 2021). When coupled with body blending, Deepfakes pose equally serious problems associated with identity theft and abuse in XR spaces. Because our XR bodies can become deeply integrated into our self-narratives/conceptions, losing control of these bodies, via XR

Deepfaked body blending, creates the possibility for users to inflict new forms of embodied trauma.

Presence, immersion, body blending, and virtually real experiences are what make XR unique, what makes our experiences with XR so fun, but these features pose a serious risk to mental health. These technologies introduce new possibilities for sexual (and other forms of) harassment that, we argue, are best dealt with proactively instead of reactively. There's a big difference between feeling present inside a medieval castle fighting off attackers and feeling present during a sexual assault. Tech author Jordan Belamire first wrote about her experience with XR sexual assault right as commercial VR headsets were becoming available:

> Outside the total immersion of the QuiVr world, this must have looked pretty funny, and definitely not real. Remember that little digression I told you about how the hundred foot drop looked so convincing? Yeah. Guess what. The virtual groping feels just as real.
> (Belamire October 20, 2016).

Harm-based concepts that were originally created to work in the context of purely physical spaces like stalking and bullying needed to adapt to the communication affordability provided by internet technology (Barak 2005). Cyberstalking and cyberbullying bear some resemblance to their physical counterparts but take into account the unique way that these forms of harassment manifest themselves online. In some cases we can readily identify harms across media. Verbal harassment, for example, is pretty much invariant in David Chalmers' sense (Chalmers 2017; Chalmers 2022).[6] Verbalized harassment remains verbalized harassment whether the audio hitting our ears is caused by vibrating vocal cords, smartphone speakers, or gaming headsets. Some forms of harassment, however, are going to require us to change our understanding of how harassment can work.

XR environments are already forcing us to reflect and revise these concepts. Take, for example, the concept of physical assault. In traditional PC-based interactions between avatars, the idea of physical assault (as opposed to other forms of harassment) has not been taken seriously. It's easy to understand how non-consensual contact between avatars is a kind of graphic harassment but can it be understood as genuinely physical harassment? We think it can (Ramirez et al. 2023). Perhaps the most obvious way is via haptics-based physical harassment. Anyone wearing a device giving them haptic feedback can be non-consensually touched by other users either due to the haptic settings of the device itself, the rules of the social XR environment (e.g., if it transmits haptic feedback to users when they come into contact with others), or because the haptic device itself has been taken over. Body blending and virtually real experience can also help explain why Belamire made a point to note that her "virtual groping feels just as real" even without haptic feedback.

Virtually real experiences are experiences that users treat, at least at the moment when they're being experienced, *as if* they were real (Ramirez 2022). This view of a simulated experience is consistent with enactivist theories of perception more generally wherein our expectations of an environment (the action affordability we expect it contains) can form a part of our perceptual experience (Nöe 2004). Virtually real experiences where our virtual bodies are touched can thus, we argue, take on a distinctly physical experience even without haptics. In one interesting study, Qingxiao Zheng and her colleagues investigated social XR interactions like this. One user reported that "when someone headpats me (well, my avatar, but to me, they're the same thing...) it gives me a tingly feeling on my head" (Zheng et al. 2022). The expectation of touch produced the sensation of touch for this user and thus, we think, a distinct form of physical harassment.

The future of social XR is a future where our concepts of harassment, sexual or otherwise, must expand to make sense of the unique action, representation, and communication opportunities these environments introduce. Mitigating these concerns, especially proactively, will require transformative approaches along several dimensions. Perhaps the most powerful way of minimizing these harms is to change our norms. XR, as we just outlined, combines elements of both physical and digital spaces, and as a result, it's unclear how our everyday social norms should apply to them. Studies on the development of social norms in XR bear out this dual nature. Consider this pair of observations from one such study:

> If one tries to be physically close to another user in social VR without asking for permission, the majority of our participants...would consider it a potential form of harassment – because it disrupts the social norm of appropriate physical distancing...In the offline world, a stranger who attempts to perform similar uninvited intimate behaviors on another stranger without consent is often considered as a harasser. In social VR, people seem to hold the same understanding.
>
> (Freeman et al. 2022, 11)

Though our participants tend to understand harassment in social VR as a simulation of harassing behaviors in the physical world (e.g., physical closeness or touch without consent), there seems to be no consensus on whether and to what degree social norms in the physical world should also be applied to social VR. These confusions further complicate why harassment in social VR occurs and how to regulate it.

(Freeman et al. 2022, 21)

The norms that we use to guide social interactions in XR, much as in physical spaces, will play the largest role in determining the boundaries between acceptable and unacceptable behavior. Because those rules are being determined *now*, it's especially important that the corporations developing these spaces draw from a representative sampling of the world's many diverse peoples and cultures. Early adopters are seldom representative of the general population.

Beyond social norms, professional organizations can do much to help mitigate forms of harassment. Organizations like the Association for Computing Machinery (ACM), the Institute of Electrical and Electronics Engineers (IEEE), and the American Psychological Association can influence the working environment of tens of thousands of professionals. Major professional organizations like these all have a code of ethics developed by their members that represent their values. Such codes need to do a better job of incorporating the potential harassment opportunities created by XR into their codes.

Lastly, existing law, and international agreements, need to be restructured to mitigate the potential for harm with social XR. Here the way forward is even murkier given the intra and inter-national nature of the problems. Throughout 2022 and 2023 the European Parliament is, for example, holding a series of consultations intended to inform future policy oriented toward "the metaverse" (European Commission 2023). Some philosophers of technology have questioned whether this is even the right sort of regulatory approach for these environments or whether social digital spaces instead "should be conceptualised and governed more like a condominium …rather than like a new frontier that can be appropriated and colonised by anybody, or like a space that belongs to no one, like the Moon" (Floridi 2021).

Conclusion

The articles in this collection prompt numerous issues for further consideration. Some of these issues seem new, while others seem to be new guises for old philosophical problems. Some of these issues appear unique to XR, while others do not. Where possible, we have suggested particular proposals here for further investigation. In highlighting these proposals, we certainly do not mean to suggest that we have exhausted the potential philosophical problems raised by XR. Rather, we think that there is an important lesson to be learned. Whether the problems are old or new, unique or general, they overwhelmingly arise from the ways that underlying technologies are overlapping and finding numerous new (and often unexpected) applications. As such, the philosophical questions being raised here are unlikely to disappear, even if current manifestations of XR technologies are only a temporary step towards new technologies.

Notes

1 By way of comparison, the most concurrently experienced game in 2022 was *Lost Ark*. At its peak it had 1.2 million concurrent users (Bankhurst, February 16, 2022).
2 See Silcox (2017) for an in depth exploration of Nozick's experience machine as a virtual world.
3 Depending on how one defines 'game', it might not even be the case that an extension is necessary. Our only point here is that literature on the philosophy of videogames and game studies may make important contributions to the philosophical exploration of virtual reality beyond virtual games.
4 The *exact* relationship Plato and Aristotle have in mind between love and the body is contentious. In Plato's *Symposium*, the character Socrates begins the search for love (understood as knowledge of Beauty) in the appreciation of distinct beautiful bodies. Aristotle, on the other hand, argues that sexual desire is, in the end, valuable only because it gives us something else that we value: affection: "intercourse then is not an end at all or is an end for the sake of receiving affection. The same goes for other appetites and arts, too" (Aristotle 1999; Sihvola 2002). So long as there is a place for physical sex in an account of love, the problems we introduce here will loom large.
5 Theories of romantic love often leave asexual relationships outside of the circle due to the (sometimes missing) component of sexual desire. While Asexuality may be captured by Aristotelian notions of affection, it's also possible that ameliorative definitions of romantic love that move away from sex may be necessary. Already romantic relationships are being critically reexamined and reanalyzed (Lindholm 1998; Nyholm, Danaher, & Earp 2022).
6 Chalmers introduces the concept of organizational invariance to explain why, in his view, some digital or simulated objects simply are what they represent themselves to be (e.g., a calendar, a calculator). In his terms, a property is organizationally invariant if it "depends only on the abstract causal organization of the underlying system…A property such as being a calculator depends only on this organization, which is also present in a simulation, so a simulated calculator is a calculator. The same reasoning explains why a virtual calculator is a calculator" (Chalmers 2017, 325).

References

Aas, S. (2021). Prosthetic embodiment. *Synthese, 198*(7): 6509–6532.

Ali, R. (2015). A new solution to the gamer's dilemma. *Ethics and Information Technology, 17*(4): 267–274.

Ali, R. (2022). The video gamer's dilemmas. *Ethics and Information Technology, 24*(2): 18.

Ali, R. (2023). The values of the virtual. *Journal of Applied Philosophy*, 40(2): 231–245.

Aristotle. (1999). *Nicomachean Ethics*, Terence Irwin (trans.), Indianapolis, IN: Hackett.

Banakou, D., Hanumanthu, P. D., & Slater, M. (2016). Virtual embodiment of white people in a Black virtual body leads to a sustained reduction in their implicit racial bias. *Frontiers in Human Neuroscience*, 10: 601. Doi: 10.3389/fnhum.2016.00601

Bankhurst, A. February 16, 2022. Lost ark has become steam's second most-played game of all time by concurrent players. *IGN*. Retrieved at https://www.ign.com/articles/lost-ark-has-become-steams-second-most-played-game-of-all-time-by-concurrent-players

Barak, A. (2005). Sexual harassment on the internet. *Science and Computer Review*, 23(1): 77–92.

Bartel, C. (2012). Resolving the gamer's dilemma. *Ethics and Information Technology*, 14: 11–16.

Bartel, C. (2022). *Video Games, Violence, and the Ethics Of Fantasy: Killing Time.* Bloomsbury Academic.

Belamire, H. (October 20, 2016). My first virtual reality groping. *Medium*. Retrieved from: https://medium.com/athena-talks/my-first-virtual-reality-sexual-assault-2330410b62ee#.8lcy2o2bh

Bellingham, H. (July 26, 2022). Cozy gaming: Why a wholesome trend became a recognized genre. *Gamesradar*. Retrieved at https://www.gamesradar.com/cozy-gaming-why-a-wholesome-trend-became-a-recognized-genre/

Ben-Ze'ev, A. (2004). *Love Online: Emotions on the Internet*. Cambridge: Cambridge University Press.

Blackwell, L., Ellison, N., Elliott-Deflo, N., & Schwartz, R. (2019). Harassment in social virtual reality: Challenges for platform governance. *Proceedings of the ACM on Human-Computer Interaction*, 3(CSCW): 1–25

Bonthuys, D. (March 30, 2023). PlayStation VR 2 is apparently selling really badly - report. *Gamespot*. Retrieved at https://www.gamespot.com/articles/playstation-vr-2-is-apparently-selling-really-badly-report/1100-6512820/

Cattien, J. (2019). Against "transracialism": Revisiting the debate. *Hypatia, 34*(4): 713–735.

Chalmers, D. J. (2017). The virtual and the real. *Disputatio*, 9(46): 309–352.

Chalmers, D. J. (2022). *Reality+: Virtual Worlds and the Problems of Philosophy.* New York: W.W. Norton & Company.

Congressional Research Service. (May 26, 2022). Military applications of extended reality. Retrieved from https://crsreports.congress.gov/product/pdf/IF/IF12010

Cook, D. (January 24, 2018). Cozy games. Lost gardens. Retrieved at: https://lost-garden.home.blog/2018/01/24/cozy-games/

de Ruiter, A. (2021). The distinct wrong of Deepfakes. *Philosophy & Technology*, 34: 1311–1332

Dembroff, Robin (2020). Beyond Binary: Genderqueer as Critical Gender Kind. *Philosophers' Imprint* 20 (9):1-23.

Dick, E. (March 4, 2021). Balancing user privacy and innovation in augmented and virtual reality. *Information Technology & Innovation Foundation*. Retrieved from https://itif.org/publications/2021/03/04/balancing-user-privacy-and-innovation-augmented-and-virtual-reality/

Ebert, R. (April 16, 2010). Video games can never be art. *RogerEbert.com*. Retrieved at: https://www.rogerebert.com/roger-ebert/video-games-can-never-be-art

emteq labs (n.d.). Measure what matters. https://www.emteqlabs.com/#data-section

European Commission. (2023). Virtual worlds (metaverses) - A vision for openness, safety and respect. Retrieved from https://ec.europa.eu/info/law/better-regulation/have-your-say/initiatives/13757-Virtual-worlds-metaverses-a-vision-for-openness-safety-and-respect_en

Floridi, L. (2021). Trump, Parler, and regulating the infosphere as our commons. *Philosophy & Technology, 34*: 1–5.

Ford, H. (June 1, 1927). New Ford car announced: Details forthcoming soon. *Ford News*. Retrieved from: https://www.thehenryford.org/collections-and-research/digital-resources/popular-topics/henry-ford-quotes

Freeman, G., Zamanifard, S., Maloney, D., & Adkins, A. (2020, April). My body, my avatar: How people perceive their avatars in social virtual reality. In *Extended Abstracts of the 2020 CHI Conference on Human Factors in Computing Systems* (pp. 1–8).

Freeman, G., Zamanifard, S., Maloney, D., & Acena, D. (2022). Disturbing the peace: Experiencing and mitigating emerging harassment in social virtual reality. *Proceedings of the ACM on Human-Computer Interaction, 6*(CSCW1): 1–30.

Gerstner, B. (October 24, 2022). Time to get fit — an Open letter from altimeter to mark Zuckerberg (and the Meta Board of Directors). *Medium*. Retrieved at: https://medium.com/@alt.cap/time-to-get-fit-an-open-letter-from-altimeter-to-mark-zuckerberg-and-the-meta-board-of-392d94e80a18

Gigliotti, C. (1995). Aesthetics of a virtual world. *Leonardo, 28*(4): 289–295.

Gladden, R., & Greenberg, E. (2015). TRANSgressive talk: An introduction to the meaning of transgracial identity. *Queer Cats Journal of LGBTQ Studies, 1*(1): 75–86.

Greener, R. (January 17, 2023). Mojo Vision cancels AR contact lens. *XR Today*. Retrieved at https://www.xrtoday.com/augmented-reality/mojo-vision-cancels-ar-contact-lens/

Hall, A. (July 28, 2022). The world's most popular social VR game is in turmoil. *Kotaku*. Retrived at https://kotaku.com/vrchat-virtual-reality-quest-anti-cheat-mods-facebook-1849341196

Hargittai, E. & Marwick, A. (2016). "What can I really do?" Explaining the privacy paradox with online apathy. *International Journal of Communication, 10*: 3737–3757.

Heinrichs, J. H. (2021). Virtual action. *Ethics and Information Technology, 23*(3), 317–330.

Heller, B. (2021). Watching androids dream of electric sheep: Immersive technology, biometric psychography, and the law. *Vanderbilt Journal of Entertainment and Technology Law, 23*(1): 1–51.

Hunting, J. (Director). (2022). *We Met in Virtual Reality*. [Film]

Incao, J. (September 20, 2018). How VR is transforming the way we train associates. *WalMart*. Retrieved from https://corporate.walmart.com/newsroom/innovation/20180920/how-vr-is-transforming-the-way-we-train-associates

Juul, J. (2001). Games telling stories. *Game Studies*, 1(1): 45.

Kugler, M. B. (2019). From identification to identity theft: Public Perceptions of biometric privacy harms. *10 UC Irvine Law Review*, 107–142.

Lindholm, C. (1998). The future of love. In de Munck, V. (ed.), *Romantic Love and Sexual Behavior: Perspectives from the Social Sciences*. Westport: Praeger.

Luck, M. (2009). The gamer's dilemma: An analysis of the arguments for the moral distinction between virtual murder and virtual paedophilia. *Ethics and Information Technology*, 11(1): 31–36

Luck, M. (2022). The grave resolution to the Gamer's Dilemma: An argument for a moral distinction between virtual murder and virtual child molestation. *Philosophia*, 50(3): 1287–1308

Madary, M., & Metzinger, T. (2016). Real virtuality: A code of ethical conduct. Recommendations for good scientific practice and the consumers of VR-technology. *Frontiers in Robotics and AI*, 3(3).

Merleau-Ponty, M. (2013). *Phenomenology of perception*. London: Routledge.

Mallon, R. (2006). 'Race': Normative, not metaphysical or semantic. *Ethics*, 116(3): 525–551

Mallon, R. (2022). What's at stake in the race debate? *The Southern Journal of Philosophy*, 60: 54–72.

Miller, M. R., Herrera, F., Jun, H., Landa, J. A., & Bailenson. J. N. (2020). Personal identifiability of user tracking data during observation of 360-degree VR video. *Scientific Reports*, Article number: 17404

Natural facial expressions privacy notice. (2022). Retrieved from: https://www.meta.com/help/quest/articles/accounts/privacy-information-and-settings/natural-facial-expressions-privacy-notice/

Nguyen, C. T. (2020). *Games: Agency as Art*. New York: Oxford University Press.

Nöe, A. (2004). *Action in Perception*. Cambridge, MA: The MIT Press.

Nozick, R. (1974). *Anarchy, State, and Utopia* (Vol. 5038). New York: Basic Books.

Nyholm, S., Danaher, J., & Earp, B. D. (2022). The technological future of love. In Grahle, A., McKeever, N. & Saunders, J. (eds.), *Philosophy of Love: In the Past, Present, and Future*. New York: Routledge.

Patridge, S. (2011). The incorrigible social meaning of video game imagery. *Ethics and Information Technology*, 13: 303–312.

Plato. (1989) *The Symposium*. Indianapolis, IN: Hackett Publishing Company.

Pritchard, T. (April 21, 2023). Apple VR/AR headset: Everything we know so far. *Tom's guide*. Retrieved at https://www.tomsguide.com/news/apple-vr-and-mixed-reality-headset-release-date-price-specs-and-leaks

Ramirez, E. J. (2020). How to (dis) solve the Gamer's Dilemma. *Ethical Theory and Moral Practice*, 23(1): 141–161.

Ramirez, E. (2022). *The Ethics of Virtual and Augmented Reality: Building Worlds*. New York: Routledge.

Ramirez, E. J., Elliott, M., & Milam, P. E. (2021). What it's like to be a _____: why it's (often) unethical to use VR as an empathy nudging tool. *Ethics and Information Technology*, 23: 527–542.

Ramirez, E. J., Jennett, S., Tan, J., Campbell, S., & Gupta, R. (2023). XR embodiment and the changing nature of sexual harassment. *Societies*, 13(2): 36.

Robson, J. (2018). The beautiful gamer?: On the aesthetics of videogame performances. In *The Aesthetics of Videogames*. Routledge: 78–94.
Roth, E. (October 15, 2022). Meta's Horizon Worlds VR platform is report-
edly struggling to keep users. *The Verge*. Retrieved at https://www.theverge.
com/2022/10/15/23405811/meta-horizon-worlds-losing-users-report
Schrader, E. (October 28, 2022). 'Traveling while black': Virtual reality exhibit
at civil rights memorial center will immerse audiences in black experiences on
the road. *Southern Law Poverty Center*. Retrieved at https://www.splcenter.org/
news/2022/10/28/traveling-while-black-virtual-reality-civil-rights-memorial-
center-black-experience
Sherman, N. (1993). Aristotle on the shared life. In Badhwar, N. K. (ed.), *Friend-
ship: A Philosophical Reader*. Ithaca and London: Cornell University Press:
91–107.
Sicart, M. (2011). *The Ethics of Computer Games*. Cambridge, MA: MIT press.
Silcox, M. (Ed.). (2017). *Experience Machines: The Philosophy of Virtual Worlds*.
Rowman & Littlefield.
Sihvola, J. (2002). Aristotle on sex and love. In Nussbaum, M. & Sihvola, J. (eds.),
*The Sleep of Reason: Erotic Experience and Sexual Ethics in Ancient Greece and
Rome*. Chicago: University of Chicago Press.
Silverman, A. M. (2015). The perils of playing blind: Problems with blindness sim-
ulation and a better way to teach about blindness. *Journal of Blindness Innova-
tion and Research*, 5(2). Retrieved from https://nfb.org/images/nfb/publications/
jbir/jbir15/jbir050201.html?sfns=mo
Smuts, A. (2005). Are video games art? *Contemporary Aesthetics (Journal Ar-
chive)*, 3(1): 6.
Solomon, R. (1988). About love: Reinventing romance for our times. New York:
Simon and Schuster.
Spiegel, J. S. (2018). The ethics of virtual reality technology: Social hazards and
public policy recommendations. *Science and Engineering Ethics*, 24: 1537–1550.
Springarn-Koff, J. (Director). (2010). *Life 2.0*. [Film].
Sunstein, C. (2015). The ethics of nudging. *Yale Journal of Regulation*, 32(2):
414–450.
Tavinor, G. (2009). *The Art of Videogames*. Malden, MA: Wiley-Blackwell.
Tavinor, G. (2021). *The Aesthetics of Virtual Reality*. New York: Routledge.
Tuvel, R. (2017). In defense of transracialism. *Hypatia*, 32(2): 263–278.
Vallor, S. (2012). Flourishing on facebook: virtue friendship: New social media.
Ethics and Information Technology, 14: 185–199.
Waszkiewicz, A., & Bakun, M. (2020). Towards the aesthetics of cozy video games.
Journal of Gaming & Virtual Worlds, 12(3): 225–240.
Wells, C. W. (2007). The road to the Model T: Culture, road conditions, and inno-
vation at the dawn of the American motor age. *Technology and Culture*, 48(2):
497–523.
Ye, J., & Yang, Y. (June 20, 2022). Tencent forms 'extended reality' unit as
metaverse race gathers steam. *Reuters*. Retrieved from https://www.reuters.com/
world/china/tencent-forms-extended-reality-unit-metaverse-race-gathers-steam-
sources-2022-06-20/

Yee, N., & Bailenson, J. (2007). The proteus effect: The effect of transformed self-representation on behavior. *Human Communication Research*, *33*(3), 1: 271–290.

Young, G. (2016). *Resolving the Gamer's Dilemma: Examining the Moral and Psychological Differences Between Virtual Murder and Virtual Paedophilia*. Springer Cham: Switzerland.

Zheng, Q., Wang, L., Ngoc Do, T., & Huang, Y. (2022). Facing the illusion and reality of safety in social VR. *CHI EA '22, Proceedings of the 1st Workshop on Novel Challenges of Safety, Security and Privacy in Extended Reality*, April 29–May 5, 2022, New Orleans, LA, USA.

Index

Note: *Italic* page numbers refer to figures and page numbers followed by "n" denote endnotes.